This page intentionally left blank.

RESTRICTED　　　　　　　　　　　　　　　　　　　　　　　AAF MANUAL 51-127-5

PILOT TRAINING MANUAL FOR
THE P-51 MUSTANG

This revised edition supersedes the original (gray cover) Pilot Training Manual for the Mustang
All copies of the latter are rescinded

Hq. Army Air Forces
Washington 25, D.C. 15 Aug 45

The use and authentication of this manual are governed by the provisions of AAF Regulation 50-17.

BY COMMAND OF GENERAL ARNOLD

Ira C. Eaker
Lieutenant General, United States Army
Deputy Commander, Army Air Forces

ADDITIONAL COPIES of this manual should be requested from: Headquarters AAF, Office of Flying Safety, Safety Education Division, Winston-Salem 1, North Carolina

INITIAL DISTRIBUTION revised edition: Headquarters AAF, 3rd Air Force, AAF Training Command

RESTRICTED

INTRODUCTION

THIS MANUAL is the text for your training as a P-51 pilot.

The Air Forces' most experienced training and supervisory personnel have collaborated to make it a complete exposition of what your pilot duties are, how each duty will be performed, and why it must be performed in the manner prescribed.

The techniques and procedures described in this book are standard and mandatory. In this respect the manual serves the dual purpose of a training checklist and a working handbook. Use it to make sure that you learn everything described herein. Use it to study and review the essential facts concerning everything taught. Such additional self-study and review will not only advance your training, but will alleviate the burden of your already overburdened instructors.

This training manual does not replace the Technical Orders for the airplane, which will always be your primary source of information concerning the P-51 so long as you fly it. This is essentially the textbook of the P-51. Used properly, it will enable you to utilize the pertinent Technical Orders to even greater advantage.

H H Arnold

COMMANDING GENERAL, ARMY AIR FORCES

HISTORY OF THE P-51

Like the Indian braves of the old southwest whose favorite in battle was the small speedy Mustang, young fighter pilots today, with their newly won wings, almost without exception want to fly the famous namesake of that sleek and powerful war horse.

And no wonder.

For the P-51 is truly a pilot's airplane. In mission after mission it has proved that it can more than hold its own against any opposition. Its speed and range are tops. It operates effectively on the deck and all the way up to 40,000 feet. In maneuverability and load-carrying capacity, it ranks with any other fighter in the world.

The P-51 was the first airplane of this war to be built entirely on the basis of combat experience. Its design was started after Hitler's Luftwaffe had begun to overwhelm Europe—after a good many lessons had been learned about modern aerial warfare from actual experience.

Mustangs were built originally for the British. Flown by RAF pilots, they saw their initial action in the summer of 1942. They were used by the British primarily for reconnaissance and rhubarb missions—for zooming in at low altitudes and strafing trains, troops, and enemy installations.

P-51s were the first American-built fighters to carry the war back across the English channel after Dunkirk. And a short time later they set another record by being the first single-engine planes of any country to penetrate Germany proper from bases in England.

So successful were the powerful little Mustangs that the United States AAF decided to adopt this American-built plane for its own.

Two improved models were built—a P-51A and an attack version known as the A-36. This attack model was equipped with bomb racks and diving brakes and was given six 50-cal. guns instead of the four 20-mm. cannon of the pursuit models. Thus, as the A-36, the Mustang became a triple-threat performer — fighter, strafer, and dive bomber. As such, it helped write aerial history in the momentous days when the Allies took Sicily and Italy.

When the need for higher altitude, longer range fighters developed so urgently, it was decided to see what the Mustang could do in these departments. So the Allison engine, with a single-speed blower, was replaced by the more powerful Rolls-Royce Merlin engine with a 2-speed blower. Along with other improvements, the prop was increased from three to four blades.

Thus was developed the P-51B and C (B if built on the west coast, C if built in Texas—they are essentially the same otherwise).

The new model proved an unquestioned success. The Nazis learned to fear it at any altitude —as high as they wanted to go. As for range, the new Mustang made it possible for the first time for fighters to escort heavy bombers all the way from Britain to Berlin.

Later, Mustangs escorted bombers all the way to Poland. And when the great triangular shuttle raids connecting England, Russia, and Italy began, P-51's were the first fighters to operate all around the continent-girdling circuit. One of the legs of this triangle was some 1600 miles long!

Last on the list, the Japanese have learned to fear the fiery breath of the Mustang. First flown by the 14th Air Force in China, P-51s have since been used with great success as long-range bomber escorts, in fighter-bomber tactics, on fighter sweeps, and for other types of tactical missions. Important victories are being rung up whenever and wherever the Japs are encountered.

The P-51D you are going to fly is a truly great airplane. Quoting an outstanding authority, who recently made a comparative analysis of all the world's aircraft: "In the single-seat fighter class, the Mustang reigns supreme." He was speaking of the B-series Mustang. The P-51D retains all the good features of its predecessor, with important added improvements. Chief among these are increased visibility for the pilot, more convenient cockpit arrangement, and heavier firepower.

Mastering the P-51, however, takes plenty of hard work. For being a first-rate fighter pilot means being not only a pilot, but a whole crew —pilot, navigator, gunner, bombardier, and radio operator—all rolled into one.

That's not a simple matter. It takes a good man to do the job right.

Remember that!

And now let's have a look at the airplane.

P-51

SERIES D AND K

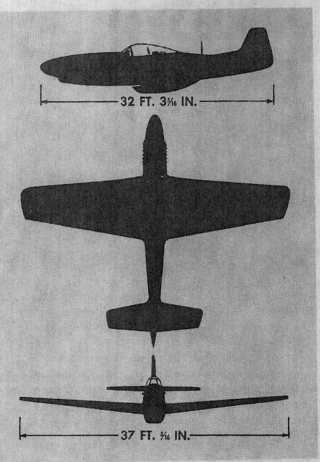

Your Mustang is a single-place, low-wing monoplane—a high-speed, long-range, low and high altitude fighter built by North American Aviation, Inc.

The fuselage is a semi-monocoque, all-metal structure, and is said to have the smallest frontal area ever placed around any high-powered liquid-cooled engine.

The all-metal wings are built in two halves which are joined at the airplane center line and are of full cantilever structure. The airfoil is of laminar-flow design which gives low drag even at high speeds.

The tail section is metal with fabric-covered elevator and rudder control surfaces. A dorsal fin gives increased lateral stability and adds to the strength of the vertical stabilizer.

The airplane is flush-riveted throughout—another factor contributing to its great speed.

CONTROLS

The ailerons, elevator, and rudder are controlled by the conventional stick and rudder pedals arrangement. All control surfaces have fiber trim tabs controllable from the cockpit by means of knobs on the left pedestal. In addition to being adjustable for trim, the rudder tab is linked so as to apply reverse boost to the rudder, improving the flight characteristics of the airplane.

The surface control lock at the base of the stick is easily operated. To lock the controls, simply bring the locking arm and the stick together while pulling out the knob on the locking arm; then release the knob. This locks all the controls. The rudders catch and are locked when moved into neutral position. To release the lock, just pull out the knob and let the locking arm spring forward out of the way.

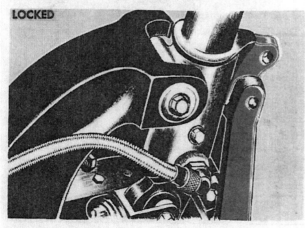

CONTROL LOCK

Note that there are two holes in the locking lug. When you use the bottom hole, the tailwheel is locked along with the controls. Using the top hole leaves the tailwheel full swiveling so the plane can be towed.

CONTROL SURFACES

The ailerons are sealed internally so that no air can pass through the opening between the aileron and the wing section. This lightens the pressures on the control stick and at the same time gives more positive action.

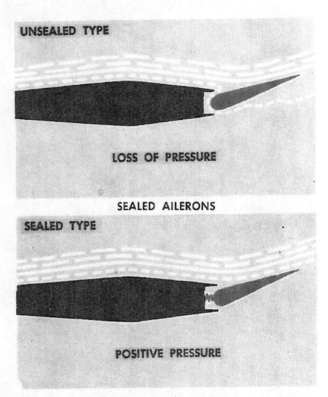

SEALED AILERONS

The flaps extend the full distance from the fuselage to the ailerons and are hydraulically operated by a control handle on the left pedestal.

Throughout their 50° range, flap position will correspond to that indicated at the control lever. Set the lever to the degree of flap desired; the flaps will automatically move to that position and stop. It takes 11-15 seconds for the flaps to go from the full up to the full down position.

COCKPIT (Front)

1. Selector Dimmer Controls
2. Remote Indicator Compass
3. Clock
4. Suction Gage
5. Manifold Pressure Gage
6. Airspeed Indicator
7. Directional Gyro
8. Artificial Horizon
9. Carburetor Air Temperature
10. Coolant Temperature
11. Tachometer
12. Altimeter
13. Bank-and-Turn Indicator
14. Rate-of-Climb Indicator
15. Oil Temperature, Fuel and Oil Pressure Gage
16. Engine Control Panel
17. Landing Gear Warning Lights
18. Parking Brake
19. Oxygen Flow Blinker
20. Oxygen Pressure Gage
21. Ignition Switch
22. Bomb and Rocket Switch
23. Chemical Release Switches
24. Cockpit Light Control
25. Gun, Camera and Sight Switch
26. Rocket Control Panel
27. Fuel Shut-off Valve
28. Fuel Selector Valve
29. Emergency Hydraulic Release

RESTRICTED

RESTRICTED

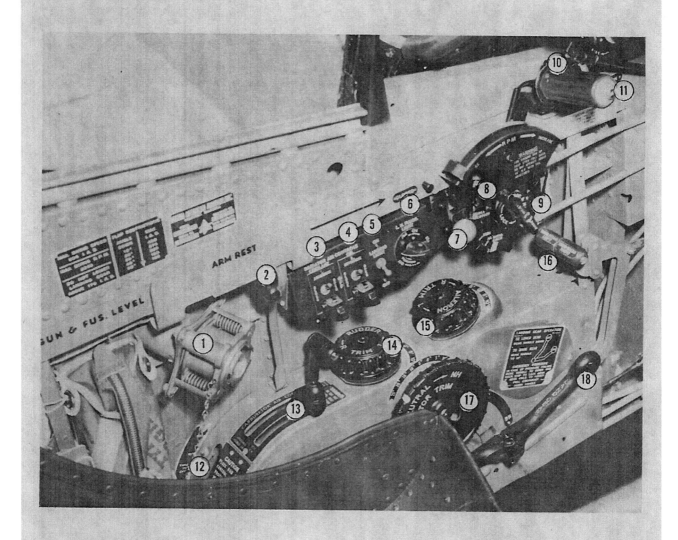

COCKPIT (Left Side)

1. Flare Pistol Opening
2. Cockpit Light
3. Coolant Radiator Air Control Switch
4. Oil Radiator Air Control Switch
5. Landing Light Switch
6. Left Fluorescent Light Switch
7. Mixture Control
8. Propeller Control
9. Throttle Quadrant Locks
10. Throttle
11. Microphone Button
12. Flap Control Handle
13. Carburetor Air Controls
14. Rudder Trim Tab Control
15. Aileron Trim Tab Control
16. Bomb Salvo Releases
17. Elevator Trim Tab Control
18. Landing Gear Control

COCKPIT (Right Side)

1. Oxygen Regulator
2. Emergency Canopy Release
3. Recognition Light Key
4. Canopy Crank and Lock
5. Circuit Breakers (under)
6. Right Fluorescent Light Switch
7. Electrical Control Panel
8. Rear Warning Radar Control Panel
9. VHF Volume Control Knob
10. VHF Control Box
11. IFF Control Panel
12. Detonator Buttons
13. Detrola Control Box
14. Cockpit Light

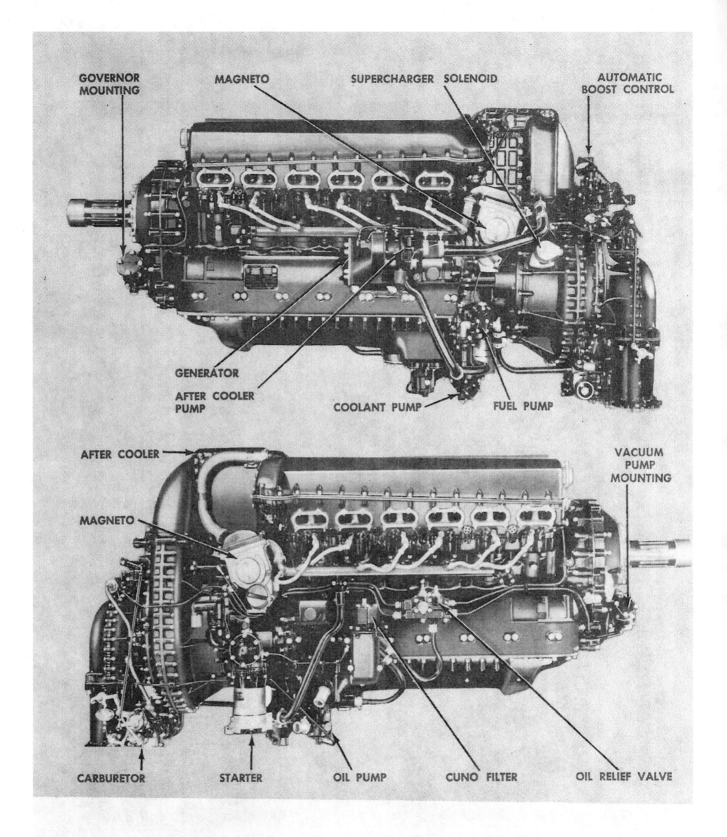

THE ENGINE

The power plant of the Mustang is a liquid-cooled, 12-cylinder, Packard-built, Rolls-Royce Merlin V-1650-3 or -7. It is equipped with an injection-type carburetor, has two-stage supercharging, and develops over 1400 hp on takeoff.

SUPERCHARGER

The engine has a two-speed, two-stage supercharger which cuts into high blower automatically. The -3 engine cuts in at 19,000 feet, the -7 engine at from 14,500 to 19,500 feet, depending on the amount of ram. The supercharger increases the blower-to-engine ratio from a low of about 6 to 1 to a high of about 8 to 1.

You can also control the supercharger manually by a switch on the instrument panel.

The switch has three positions—AUTOMATIC, LOW and HIGH.

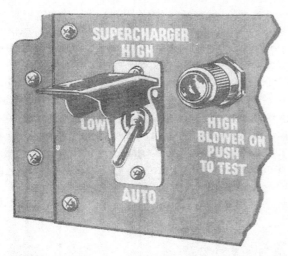

For all normal operations, keep the switch in AUTOMATIC. In this position the supercharger is controlled by an aneroid-type pressure switch, which automatically cuts the unit into high or low blower as required. This switch is so adjusted that it cuts the unit back into low blower approximately 1500 feet under the altitude at which it cuts into high blower. This prevents the high blower from going on and off repeatedly with slight changes in altitude at about the point where the high blower cuts in.

If the aneroid switch fails, the supercharger automatically returns to low blower.

The LOW position on the manual switch on the instrument panel makes it possible to operate the supercharger in low blower at high altitudes. This gives you better range at high altitudes—which, of course, is important on long-range flights.

The HIGH position on the manual switch makes it possible to test the high blower on the ground. You have to hold the switch in the HIGH position, however, because it is spring-loaded and flips back to the LOW position as soon as you let go.

An amber jewel indicator light next to the manual switch on the instrument panel goes on when the supercharger is in high blower.

CARBURETOR

The engine has an injection-type carburetor and an automatic manifold pressure regulator. With this automatic regulator, you don't have to jockey the throttle to maintain a constant manifold pressure in the high-speed range as you climb or let down. All you have to do is select the desired pressure by setting the throttle lever, and the pressure regulator does the rest. It compensates automatically for the difference in air density at different altitudes by gradually opening the carburetor butterfly valve as you climb and smoothly closing it as you let down.

On later airplanes, the automatic regulator covers practically your entire operating range, going into action whenever you use more than 20" of manifold pressure. Airplanes equipped with this type regulator can be distinguished by the START position plate on the throttle quadrant. In earlier airplanes the manifold pressure regulator is effective only at pressures in excess of 41".

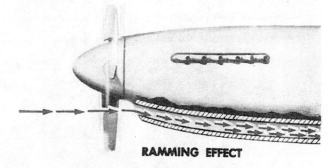

RAMMING EFFECT

Carburetor air comes through a long carburetor air scoop directly under the engine. The plane's motion forces the air at high speed (or rams it) directly into the carburetor. This is called ram air.

If the scoop becomes obstructed by ice or foreign matter, a door in the air duct opens automatically to admit hot air from the engine compartment to the carburetor.

Ordinarily you will always use ram air, but, in the event of extreme icing or dust conditions, the carburetor air controls on the left cockpit pedestal allow you to select either unrammed filtered or, in later airplanes, unrammed hot air for operation. In order to obtain hot air, the hot air control must be at HOT and the cold air control at UNRAMMED FILTERED AIR. If the cold air control is in RAM AIR position, the hot air control will be ineffective.

Don't use hot air above 12,000 feet. At high altitudes its use will disturb the carburetor's altitude compensation, and may cause too lean a mixture.

WAR EMERGENCY POWER

In order to give your engine an extra burst of power should you get into an extremely tight situation, move the throttle full forward past the gate stop by the quadrant, breaking the safety wire. The engine will then be opened up to its absolute limit, and will give you about 6" of manifold pressure in excess of the normal full throttle setting of 61" (with mixture control at RUN or AUTO RICH and prop set for 3000 rpm).

This throttle reserve is called war emergency power, and should be used only in extreme situations. If you use it for more than 5 minutes at a time you'll risk damaging vital parts of the engine. In training, therefore, the throttle must never be moved beyond the gate stop.

Whenever you do use war emergency power, be sure to note the length of time on Form 1A, and also report it to the crew chief or engineering officer so that a record can be kept and the engine inspected before the airplane is flown again. The engine must be removed for a complete knock-down inspection after 5 hours.

On some early airplanes, war emergency power is not incorporated into the throttle, but is obtained by pulling a boost control lever, on the panel forward of the control quadrant. If your airplane is so equipped, remember that there is no point in pulling war emergency power until you have opened the throttle all the way. In other words, don't use the boost control lever to increase your power when you can get the same result by opening the throttle.

Remember too, that there's nothing to be gained by using war emergency power below 5000 feet. Up to that altitude the throttle alone gives you more than enough power to exceed the operating limits of the engine.

THROTTLE QUADRANT

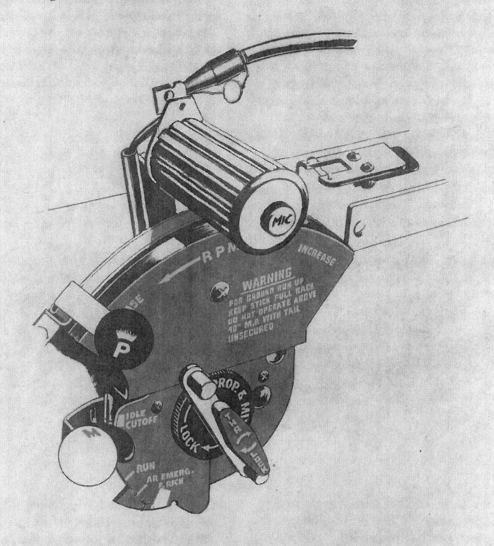

There are two types of throttle quadrants, the principal difference depending upon whether the airplane is equipped with a single- or two-position carburetor. On later airplanes with single-position carburetors, the mixture control has the following settings: IDLE CUT-OFF, RUN, and EMERGENCY FULL RICH. These carburetors are fully automatic and the normal operating position is RUN. The EMERGENCY FULL RICH position is for use in case the carburetor fails to function properly in RUN. To place the mixture control in the EMERGENCY FULL RICH setting, a spring latch on the lever must be pressed with the thumb as the control is moved past a break-through seal. On earlier airplanes with two-position carburetors, the mixture control positions are IDLE CUT-OFF, AUTO LEAN, and AUTO RICH.

If your airplane has the latest type manifold pressure regulator, you'll find a START position designated on the throttle quadrant, necessary because this pressure regulator requires a 2-inch throttle opening for starting the engine, whereas in earlier airplanes the throttle is to be opened only one inch.

Some throttles are fitted with special twist grips, used in operation of the K-14A gunsights installed in later airplanes.

The quadrants have two friction-lock adjusting knobs. One adjusts the friction of the propeller and mixture control levers, the other the throttle control lever.

PROPELLER

The P-51D propeller is a Hamilton Standard, four-blade, hydraulic, constant-speed prop with a diameter of 11 feet 2 inches and a blade angle range of 42°. As is the case with all single-engine aircraft, the prop cannot be feathered. You control propeller rpm manually by a single lever on the throttle quadrant.

P-51K airplanes are provided with Aeroproducts propellers. These, too, are four-blade, hydraulically operated, constant-speed propellers, diameter 11 feet and with a pitch range of 35°.

The P-51D and P-51K series airplanes are distinguished from each other solely by the prop installation. And, although the Hamilton Standard and Aeroproducts propellers are radically different in construction, they are identical in operation.

LANDING GEAR

The main gear and tailwheel are fully retractable, and are controlled hydraulically by a single lever on the left pedestal.

Note in the accompanying illustration that this lever is locked in position; you must pull it inboard before pulling it up to raise the gear. Note also that in lowering the gear, the landing gear lever must be all the way down **and locked**. If it isn't locked, the lever may creep back up.

Do not raise the control lever when the airplane is on the ground. There is no safety downlock on a P-51D, and the gear will retract as soon as you start taxiing.

The tailwheel is both steerable and full swiveling. It is steerable 6° right or left with the rudder. The tailwheel lock is different from that of most other planes—it is operated by the control stick. When the stick is in neutral position or pulled back, the tailwheel is locked and steerable. When you push the stick full forward, the tailwheel is unlocked and full swiveling.

PULL INBOARD TO UNLOCK

The tailwheel drops almost instantly when you push the landing gear lever to the DOWN position. The main gear takes 10 to 15 seconds to move into position. You can definitely feel the gear lock into place when it is lowered.

There'll be occasions, such as go-arounds, when you'll want to raise the landing gear immediately after lowering it. In doing so, always wait until the gear is down and locked before pulling up the control lever. If you attempt to raise the gear while it's still on the way down, you'll succeed only in damaging it or the fairing doors.

You can release the landing gear in an emergency by means of a red handle just above the hydraulic pressure gage. Pulling this red handle releases the pressure in the hydraulic lines. This allows the gear to drop of its own weight when the landing gear lever is in the DOWN position. The landing gear lever must be in the DOWN position—all the way down—or the mechanical locks which hold the gear in place are not unlocked.

Note that the red handle releases the pressure in the hydraulic lines. Therefore, if you want to operate the flaps or the fairing doors after you have dropped the gear by means of the red handle, you must push the handle back to its original position. If you leave it out—all or even part way out—you won't have any hydraulic pressure to operate the flaps.

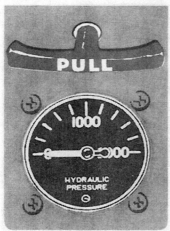

EMERGENCY HYDRAULIC RELEASE

If the gear is not down and locked when you come in for a landing, you'll be warned by a combination of red and green warning lights below the instrument panel and a horn aft of the seat. The horn sounds only when the landing gear is up and locked, while the throttle is retarded below the minimum cruise condition, and may be silenced by the cut-out switch on the front switch panel. This switch automatically resets when the throttle is advanced.

Horn Blowing

THROTTLE . . . RETARDED
DOORS . . . . CLOSED
GEAR. . . . . UP AND LOCKED OR DOWN AND UNLOCKED

THROTTLE . . . ANY POSITION
DOORS . . . . OPEN
GEAR. . . . . DOWN AND UNLOCKED OR UP AND LOCKED

THROTTLE . . . ANY POSITION
DOORS . . . . ANY POSITION
GEAR. . . . . DOWN AND LOCKED

THROTTLE . . . ADVANCED
DOORS . . . . CLOSED
GEAR. . . . . UP AND LOCKED

Operation of the lights and horn is explained in the accompanying illustration. The lights are equipped with adjustable dimmer masks, and are of the push-to-test type, so that you can readily check to see if the bulbs are okay.

BRAKES

The brakes are hydraulic, of the disc type. The usual toe action on the rudder pedals controls each brake individually.

After takeoff, never brake the wheels to stop them from turning. If the brakes are hot from excessive ground use, they are likely to freeze. The design of the gear and the wheel wells is such that under normal conditions the turning of the wheels has no harmful effect even after they have been retracted into the wheel wells.

The parking brake handle is just below the center of the instrument panel. You operate the parking brakes in the conventional manner:

1. Hold the brakes;
2. Pull the parking brake handle out;
3. Release the pressure on the brake;
4. Then release the parking brake handle.

To release the parking brakes, simply push down on the foot pedals.

Caution: Never set the parking brakes when the brakes are hot; the discs may freeze.

HYDRAULIC SYSTEM

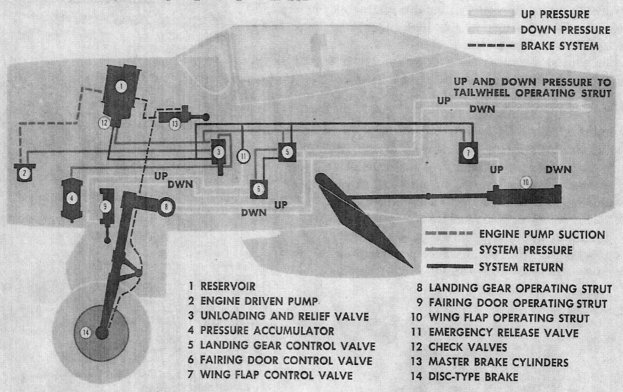

1 RESERVOIR
2 ENGINE DRIVEN PUMP
3 UNLOADING AND RELIEF VALVE
4 PRESSURE ACCUMULATOR
5 LANDING GEAR CONTROL VALVE
6 FAIRING DOOR CONTROL VALVE
7 WING FLAP CONTROL VALVE
8 LANDING GEAR OPERATING STRUT
9 FAIRING DOOR OPERATING STRUT
10 WING FLAP OPERATING STRUT
11 EMERGENCY RELEASE VALVE
12 CHECK VALVES
13 MASTER BRAKE CYLINDERS
14 DISC-TYPE BRAKE

The hydraulic system is extremely simple and almost foolproof in its operation. It has two separate parts. One part of the system operates under pressure from a pump which is driven directly off the engine. This part operates the landing gear and the wing flaps. This "power" part of the system operates at a pressure of 1000 pounds per square inch (psi) while the engine is running.

The second part of the system works the brakes only. It is operated by the foot pressure of the pilot.

The only connection between the two parts of the system is that they both receive their supply of fluid from the same tank. However, the tank is so designed that even if all the hydraulic fluid from the power part of the system is lost, there still is enough fluid to operate the brakes. So even if you lose the hydraulic pressure in your landing gear and flaps, you still can operate your brakes.

ELECTRICAL SYSTEM

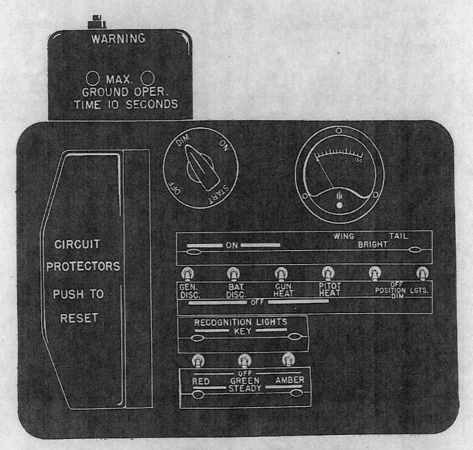

ELECTRICAL CONTROL PANEL

The electrical system is a 24-volt, direct-current system which provides power for operating the booster pumps, starter, radios, guns, the various electric lights, the bomb racks, and the coolant and oil radiator controls.

The electrical system runs off the battery until the engine reaches 1500-1700 rpm, when the generator is cut in by the voltage regulator. Power for the electrical system then is supplied by the generator. To prevent any damage to the electrical system from overload, circuit breakers are used. These eliminate the use of fuses and allow you to re-set broken circuits while still in flight.

The circuit breaker re-set buttons are on the right switch panel. On late models all the buttons can be re-set at once by means of one bar plate across the switches. All you have to do is simply bump this plate to restore the circuits.

An ammeter is on the same panel as the circuit breaker switches. This ammeter shows you how much current is flowing from the generator and also shows whether or not the generator has cut in at 1500-1700 rpm as it should.

The battery is just behind the pilot's armor plate in the radio compartment. The battery and generator disconnect switches are on the panel with the circuit breaker switches.

The lights of the electrical system include cockpit and gunsight lights, one powerful sealed-beam landing light in the left wheel well, recognition lights, and standard navigation lights on the wingtips and on the rudder.

Except for the booster coil, which is used only in starting, the ignition system is completely independent of the electrical system, and will continue to function normally in case of electrical system failure. Ignition power is supplied by the magnetos; the switch is on the front switch panel.

FUEL SYSTEM

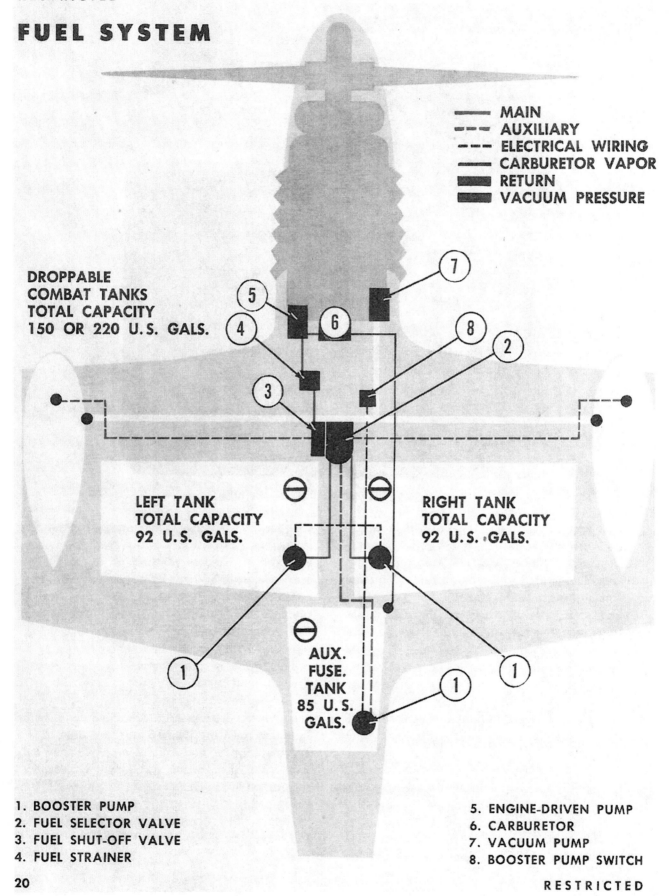

The Mustang has two main fuel tanks, one in each wing. They are self-sealing, and have a capacity of 92 gallons each. An auxiliary 85-gallon self-sealing tank is installed in the fuselage, aft of the cockpit. There is also provision for carrying two droppable combat tanks on the bomb racks. These are available in 75-gallon and 110-gallon sizes; normally you'll use only drop tanks of 75-gallon capacity, since the extra weight of the larger tanks imposes near-limit loads on the wings and bomb racks. Droppable tanks are not self-sealing. Making them so would add unnecessary weight, for it has been found that shot-up external tanks will not remain afire when the airplane is in flight.

Total fuel capacity of the airplane, with 110-gallon droppable tanks, is 489 gallons. Normally, Grade 100/130 fuel is used, and the consumption under selected cruising conditions varies considerably. To cruise the Mustang scientifically, plan your flight in accordance with the Flight Operation Instruction Charts included at the end of this manual.

Fuel is forced to the carburetor by an engine driven pump at a normal operating pressure of 16-19 psi. In addition, there is an electrically powered booster pump in each internal tank. These booster pumps prevent vapor lock at high altitudes, assure sufficient fuel supply under all flight conditions, and, in case of engine driven pump failure, will provide enough fuel to the carburetor for normal engine operation.

The three booster pumps are controlled by a single switch on the front switch panel. You simply turn this switch to ON; then, turning the fuel selector valve from one position to another automatically shuts off the booster pump in the tank just used and starts the pump in the tank selected.

On some of the earlier P-51D's, the booster pump switches have three positions—EMERGENCY, OFF, and NORMAL. In the NORMAL position the booster pump supplies fuel to the engine driven pump at a pressure of 8-10 psi. EMERGENCY increases this pressure to about 19 psi.

Use the EMERGENCY position in takeoffs and landings and in the event of engine-driven pump failure.

Fuel is boosted from the droppable tanks by means of tank pressurizing. Pressure is supplied by venting the tanks to the vacuum pump exhaust, and is maintained at a constant 5 psi by an automatic relief valve.

Fuel gages for the wing tanks are on the floor, to the left and right of your seat. The gage for the fuselage tank is mounted on the upper left side of the tank itself; you'll have to look over your left shoulder to read it. There are no gages for the droppable tanks.

FUSELAGE TANK GAGE

The fuel selector control is on the floor of the cockpit, just in front of the stick. As you rotate the valve handle you'll notice a definite snap as each tank position is reached. Be sure you feel this snap—it's your guarantee that the valve is properly set.

RESTRICTED

Whenever the engine is running, vaporized fuel is returned from the carburetor to one of the fuel tanks through the vapor return vent line. On later airplanes, this line is led to the fuselage tank, while on some of the earlier P-51D's the vapor return is to the left wing tank. Find out to which tank the vapor return line is connected by asking your engineering officer, and use fuel from that tank first to allow space for the return. Also, check the tank occasionally in flight to make sure that it isn't completely full. If you don't do this, the recovered gas, which may amount to as much as 10 gallons per hour depending on operating conditions, will be lost through the overflow pipe.

The electric primer installed on later airplanes is controlled by a momentary toggle switch next to the starter switch. Earlier series planes have a hand primer on the right side of the instrument panel. One second's operation of the electric primer is about equivalent to one stroke of the hand primer.

> **CAUTION**
>
> When changing tanks, don't stop the selector valve at an empty tank position, or at a droppable tank position if you have no droppable tank. If you should accidentally do this, or if you run a tank completely dry, your engine will fail, and you must act immediately as follows:
>
> 1. Turn the fuel selector to a full tank;
> 2. Make sure that the booster pump switch is ON;
> 3. As your engine takes hold, adjust the throttle setting as required.

Note In later planes the vapor return goes to the fuselage tank instead of the left wing tank. With these planes, use the fuselage tank first. However, the only way of finding out to which tank the return is connected is to ask the engineering officer.

OIL SYSTEM

The oil system includes a tank, located just forward of the firewall, and a radiator in the air scoop under the fuselage.

The tank is a hopper type—that is, it is designed with hoppers or compartments which facilitate quick warm-up and also make it possible to fly the airplane in abnormal positions, with little oil in the system.

With this tank you can fly the P-51 in any attitude when the tank is full. Or you can put it into a vertical climb or dive when the tank is

INVERTED FLYING LIMITED TO 10 SECONDS

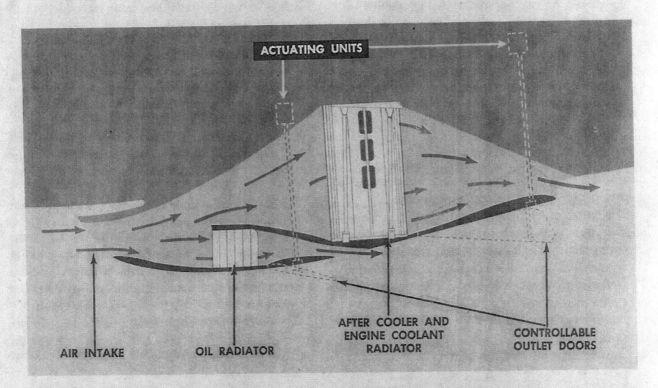

only ¼ full and still get proper lubrication. However, when the plane is in inverted flight, the oil pressure falls off because no oil reaches the scavenger pump. For that reason you must limit inverted flying to 10 seconds—which is plenty of time for any normal or combat maneuvers.

An outlet door on the bottom of the air scoop controls the oil temperature. Under ordinary circumstances this door operates automatically. However, if you want to operate it manually—when you're running your engine on the ground, for example, or in case the automatic regulator fails in the air—you can do so by means of an electrical switch on the left side of the cockpit.

This switch has three positions: AUTOMATIC, OPEN, and CLOSE. You can stop the door at any position by holding the toggle switch in the OPEN or CLOSE position for the necessary length of time, then returning the switch to neutral.

The oil system has standard AAF oil dilution equipment. This allows you to thin the oil with gasoline to make the engine easier to start in temperatures below 40°F.

RESTRICTED

OIL DILUTION SWITCH

Operation of the oil dilution equipment is simple. Allow the engine to idle, coolant flaps open, until the oil temperature drops to 50°C or less. Then, before stopping the engine, use the dilution switch on the pilot's switch panel. This dilutes the oil until you are ready to start the engine again. Once the engine warms up, the gasoline in the oil is quickly evaporated.

If the engine temperature is high, stop the engine and allow it to cool to an oil temperature of about 40°C. Then start it again, and immediately dilute the oil as explained above.

Two minutes of oil dilution is sufficient for any temperature down to 10°F. When starting in temperatures lower than that, heat is sometimes applied to the engine and oil system. Therefore, no fixed oil dilution time can be given in this manual; you'll be specially instructed in accordance with local operating conditions.

Total capacity of the oil system is 21 gallons.

COOLANT SYSTEM

With the radiators located in the big airscoop aft of the cockpit under the fuselage, the cooling of the P-51 engine is quite different from that of any other airplane.

The engine is cooled by liquid in two separate cooling systems. The first system cools the engine proper, the second (called the after-cooling system) cools the supercharger fuel-air mixture. Each performs a separate function and the systems are not connected in any way. They both pass through a single large radiator, but in different compartments.

The engine coolant system is a high-pressure system (30 psi) and its capacity is 16½ gallons. Operating pressure of the after-cooling system is lower (20 psi), and its capacity is 5 gallons.

The coolant used is a mixture of ethylene glycol and water, treated with "NaMBT" (a corrosion preventive). There are two types— type D for normal use, which consists of 30% glycol and 70% water, and type C for winter use (below 10°F), which consists of 70% glycol and 30% water.

An air outlet door at the rear of the air scoop controls the temperature of the coolant. This door operates similarly to that of the oil cooler. Normally, it works automatically, but you can control it manually by means of a switch on the left side of the cockpit, next to the oil cooler door control switch. Both switches are on the radiator air control panel.

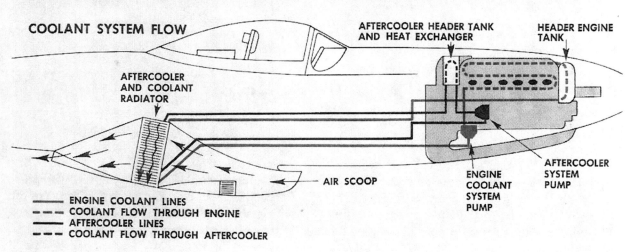

RESTRICTED

THE CANOPY

CANOPY CRANK AND EMERGENCY RELEASE INSIDE

The cockpit enclosure is of the half-teardrop type; it consists of an armor glass windshield and a sliding canopy formed from a single piece of transparent plastic. The canopy is designed to give you the best possible vision in all directions, since obstructions above, at the sides, and to the rear have been eliminated.

You get into the cockpit from the left side. To help you up on the wing, there is a handhold in the left side of the fuselage. You can step on the fairing in getting up on the wing, but be careful that you **don't** step on the flaps.

To open the canopy from the outside, push in on the spring-loaded button at the right forward side of the canopy, and slide the canopy aft.

You control the canopy from within by means of a hand crank. Depressing the latch control on the crank handle unlocks the canopy, after which you can turn the crank to slide the canopy open or closed. Releasing the latch control locks the canopy in any position.

To warn you against taking off without having the canopy properly secured to the airplane, there are two red indicator pins, one at each side of the canopy. If these pins are visible the canopy isn't properly locked.

Never take off if you can see the pins. If you do, your canopy will blow off.

EMERGENCY RELEASE INDICATOR

The emergency release for the canopy is the long red handle on your right, above the oxygen controls. When you pull this handle, the entire canopy flies off. The handle is safetied with light safety wire.

RESTRICTED

COCKPIT

The cockpits of fighter-type airplanes are generally pretty cramped, and that of the Mustang is no exception. Concentration of numerous instruments and controls into a small space is unavoidable. In the case of the P-51D, the controls are simplified, and their grouping has been planned to give you the greatest possible efficiency. As fighter airplanes go, the cockpit is comparatively comfortable.

The cockpit can be both heated and ventilated. Cold air is fed into the cockpit through a small scoop located between the fuselage and the big air scoop. Warm air is fed into the cockpit from inside the scoop just back of the radiator. Warm air from this source also serves to defrost the windshield.

The controls for regulating cold and warm air and the defroster are on the floor of the cockpit, around the seat, as shown in the accompanying illustration.

The pilot's seat is designed to accommodate either a seat-type or a back-pack parachute. The back cushion is kapok-filled and can be used as a life preserver. The seat is adjustable vertically; you'll find the lock on your right.

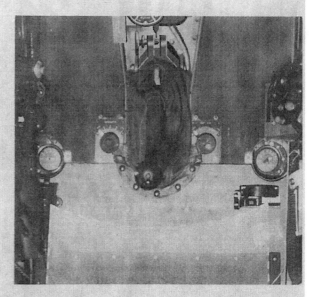

No fore-and-aft adjustment is possible.

Your comfort on long flights will be increased by a small, folding arm rest on the left side of the cockpit.

A standard safety belt and shoulder harness are provided. There is a lever on the left side of the seat for relaxing the tension on the shoulder harness. This permits you to lean forward whenever necessary—for example, to look out of the canopy in taxiing.

Use That Harness! Don't think your shoulder harness is just another troublesome gadget. If you ever have to make a forced landing, or are stopped suddenly, it may save your life. Many a pilot who has had to use the shoulder harness in an emergency now swears by it and not at it.

INSTRUMENTS

The instruments in the P-51D cockpit are shown above and on the following pages. Most of the instruments are mounted on the instrument panel, flight instruments being grouped at the center and to the left, engine instruments to the right. Exceptions are the hydraulic pressure gage, which is below the pilot's switch panel; the fuel gages, on the floor and aft of the cockpit; and the ammeter on the electrical switch and circuit breaker panel.

The instruments can be classified into four general groups, as follows:

1 VACUUM SYSTEM INSTRUMENTS

These are operated by a vacuum pump driven off the engine and include:
- the flight indicator,
- the bank-and-turn indicator,
- the directional gyro, and
- the suction gage.

The suction gage shows whether the vacuum pump is providing proper vacuum for the system. If this gage reads more than 4.25 or less than 3.75, you know that the vacuum instruments are not functioning properly and are not giving reliable readings. Normal suction reading is 4.00.

2 PITOT STATIC SYSTEM INSTRUMENTS

These instruments are operated by pressure or static air from the pitot tube, which is under the right wing, and from the static plates on the fuselage skin. They include:
- the airspeed indicator,
- the altimeter, and
- the rate-of-climb indicator.

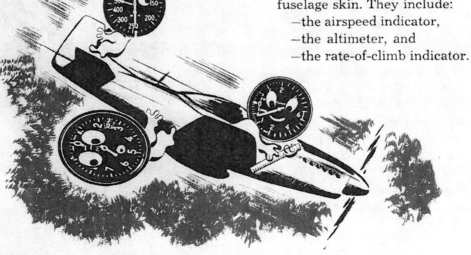

3 ENGINE INSTRUMENTS

The engine instruments include:
- the manifold pressure indicator,
- the tachometer,
- the carburetor air temperature indicator,
- the coolant temperature gage, and
- the engine gage.

The engine gage consists of three instruments in one—showing oil temperature, oil pressure, and fuel pressure.

4 MISCELLANEOUS INSTRUMENTS

Miscellaneous instruments include:
- the remote indicator compass,
- the hydraulic pressure gage,
- the oxygen pressure gage,
- the fuel gages,
- the ammeter, and
- the clock.

Note: The engine instruments are grouped on the right side of the instrument panel, the flight instruments in the center, and the miscellaneous instruments at the left.

RESTRICTED

You are already familiar with most of the instruments in the P-51, having used them in other aircraft. However, two of the instruments are probably new to you—the remote indicator compass, which is of new design, and the electrical tachometer, which differs in its operation from the mechanical type tachometer.

The remote indicator compass used on P-51D's replaces the conventional magnetic compass, although on late series airplanes a direct-reading, standby magnetic compass is also provided. The remote compass unit is in the left wingtip and transmits its readings electrically to an indicator on the instrument panel.

This type of compass doesn't float around and fluctuate when the plane maneuvers, and it gives you all the advantages of the directional gyro without the precession inevitable in the directional gyro. Both are provided, however, the directional gyro being available in an emergency should the electrical system fail.

The electrical-type tachometer operates directly from a small generator driven by the engine. The faster the engine turns over, the greater the output of this generator. The tachometer measures this generator output in terms of rpm. It operates independently of the main electrical system.

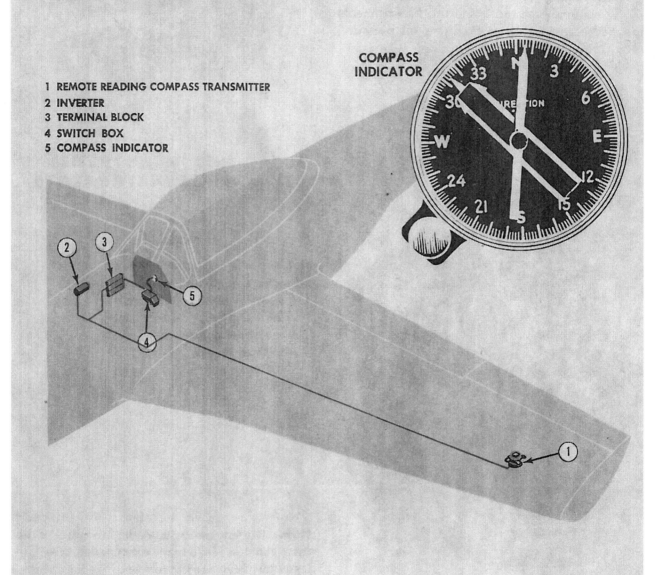

1 REMOTE READING COMPASS TRANSMITTER
2 INVERTER
3 TERMINAL BLOCK
4 SWITCH BOX
5 COMPASS INDICATOR

COMPASS INDICATOR

RADIO EQUIPMENT

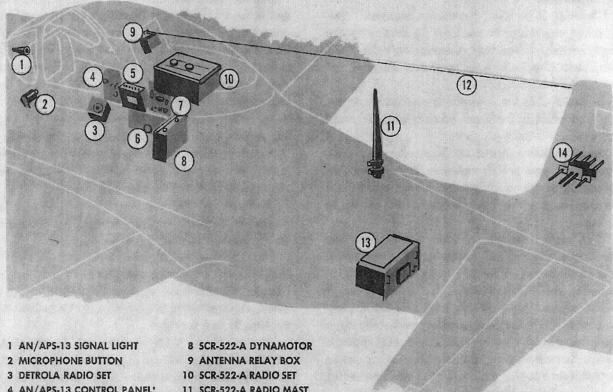

1 AN/APS-13 SIGNAL LIGHT
2 MICROPHONE BUTTON
3 DETROLA RADIO SET
4 AN/APS-13 CONTROL PANEL'
5 SCR-522-A CONTROL PANEL
6 AN/APS-13 SIGNAL BELL
7 IFF CONTROL PANEL
8 SCR-522-A DYNAMOTOR
9 ANTENNA RELAY BOX
10 SCR-522-A RADIO SET
11 SCR-522-A RADIO MAST
12 DETROLA ANTENNA
13 AN/APS-13 RADIO SET
14 AN/APS-13 ANTENNA

The radio equipment in the latest P-51D consists of a VHF (Very High Frequency) transmitter and receiver, a Detrola receiver, an AN/APS-13 rear-warning radio set, and an IFF (Identification, Friend or Foe) unit.

All radio equipment is in the fuselage, aft of the cockpit. Controls are in the cockpit, and are grouped at your right. Each set has its own antenna arrangement; the VHF antenna mast extends vertically above the fuselage aft of the cockpit, the Detrola wire antenna runs from the back armor plate to the top of the fin, the AN/APS-13 antenna rods extend horizontally from the sides of the fin, and the IFF antennae project from the undersides of the wings.

The VHF Set

You operate the VHF set by means of a control box which has five push buttons—an OFF switch and A, B, C, and D switches for selec-

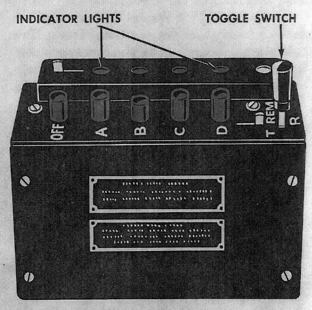

VHF CONTROL BOX

tion of four different frequencies. These frequencies are crystal controlled and, therefore, cannot be adjusted in flight. You can transmit and receive on only one channel at a time.

Channel A is for communication with CAA radio ranges.

Channel B is the "American common" frequency; use it for contacting all towers in the continental United States equipped with VHF facilities. You will also use this channel for emergency homings.

Channel C is for interplane communications.

Channel D is the local homing channel used for practice homings.

Small colored lights alongside the four buttons show you which channel is in operation. These lights have a dimmer device which can be moved over the lights to dim them for night flying.

The toggle switch on the control box has three positions—REM (remote), T (transmit), and R (receive). This switch is usually safety-wired in the REM position.

When the toggle switch is in the REM position, the VHF set is controlled remotely by a push button on the throttle. When this throttle button is pushed in, you are transmitting; when it is out in normal position, you are receiving. Under ordinary circumstances you use the remote control and the throttle button.

A tiny white light besides the toggle switch shows you whether you are transmitting. This light stays on except while you're transmitting. If it doesn't go out when you press the throttle button you know that your transmitting equipment is not working.

On the other hand, the transmitter light may go out and stay out, indicating that your transmitter is on, because of either a stuck or shorted throttle switch, or a jammed relay. In either case, try to correct the condition by operating the throttle switch a few times, and turning the main radio switch off and on. If this doesn't help, break the safety wire and move the toggle switch from REM to R or T, as required.

But remember, don't leave your transmitter on. If you do, the carrier wave will jam the channel so that no one else can use it.

You control the volume of the VHF set by means of a knob on the AN/APS-13 control panel located at your right side.

The chief advantage of VHF equipment is that it is not affected nearly as much by atmospheric interference as low-frequency equipment. Accordingly, it provides much better reception in bad weather.

The range of VHF equipment is normally about 200 miles at an altitude of 20,000 feet. Altitude and terrain determine range, because VHF transmission travels only in straight lines. In general, operation improves with increase in altitude.

To contact a VHF tower you must maintain an unobstructed line between the tower and your antenna. Remember, if the tower is below your horizon, if mountains, tall buildings, or

Remember: For VHF contact, you must maintain a clear line between plane and tower. VHF will not penetrate or travel around obstructions.

parts of your airplane are in the way, your transmission and reception will be blocked out. And here's a tip—if the tower you're contacting is close by, make a big turn around it, keeping the tower on the inside of the turn.

Don't forget that if you want to make contact with a distant VHF station you must get up high, above all obstructions. At 90 miles, for example, you must be up around 7000 feet; at 120 miles, about 10,000 feet.

The Detrola

The Detrola is a low-frequency receiver. It operates between 200 and 400 kc, which covers the transmission band for towers and range stations throughout the United States.

Operation of the Detrola is simple. It has only two controls—a station selector knob and a combination ON-OFF switch and volume control.

DETROLA CONTROL BOX

Although your headset plugs into the VHF equipment, you don't have to change your headset connection to use the Detrola. The Detrola is interconnected with the VHF equipment so that both units operate through one headset.

You can operate the Detrola and the VHF set independently or both at once. By turning on both sets you pick up not only the Detrola reception but also whatever is in the air on the particular VHF channel you are using.

Don't expect extreme long-range reception from the Detrola. When flying at altitudes of approximately 8000 feet or less, the range of your Detrola will normally be limited to about 50 miles, increasing with altitude to about 75 miles. That's about the maximum you can depend on, although under favorable conditions longer range reception may be obtained.

Remember, the Detrola is a receiver only. You **can't** talk to towers and range stations with this equipment. Your only means of talking to towers and range stations is the VHF set. That is the advantage of being able to use both sets at once—you can receive on one and transmit or receive on the other.

Rear Warning Radio

The AN/APS-13 unit is a light weight radar set that warns you, by means of an indicator light and a bell, of the presence or approach of other aircraft to your rear.

The red jewel indicator light is mounted on the left side of the instrument cowl, the bell is to the left of the seat. Control switches are provided on the panel at your right.

The equipment is automatic and is turned on and off by a toggle switch. There's also a check-

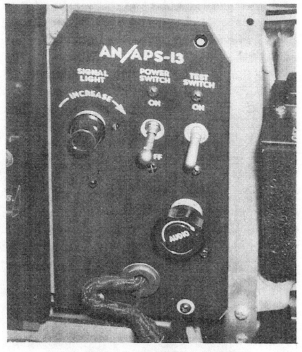

AN/APS-13 CONTROL PANEL

ing switch to test the operation of the light and bell, and a rheostat for controlling the intensity of the indicator light.

The IFF Set

The IFF set is an identification device for use in combat zones. Its operation is simple, so far as the pilot is concerned. It is automatic; all you have to do is turn it on and off with a toggle switch.

The IFF set has a detonator to destroy vital parts of the equipment if you have to abandon the airplane over unfriendly territory. The detonator is set off by pressing two push button switches which are in a box on the right side of the cockpit. Both buttons must be pressed simultaneously.

The IFF set also has an impact switch to set off the detonator in a crash landing. Thus, the equipment can be destroyed even if you have to bail out in a hurry and don't have time to press the two buttons.

RADIO NAVIGATION

To assist you in navigation, you have at your disposal a Detrola and a VHF set, which can be used independently or together. Using this equipment with regular radio navigation facilities, you should have no excuse for ever getting lost in this country, where radio facilities are unlimited.

You should be thoroughly familiar with all the aids available in order to make full use of them. Unless you are, you may find yourself helpless in a situation which requires the particular equipment with which you are **not** familiar.

Too often pilots have a tendency to concentrate on only one facility. They use their VHF set, for example, almost exclusively, neglecting their Detrola. The danger of such faulty practice is obvious, so be as thoroughly experienced as possible with all the equipment at your command.

HOMING

Homing facilities are usually available on two of the VHF channels, B and D.

The detonator destroys the equipment internally, and the explosion won't harm you or the airplane.

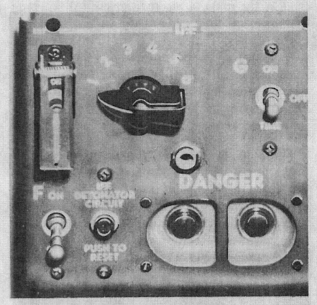

IFF CONTROL PANEL

If you ever get into an **emergency** situation and require a homing, remember these two things:

1. State definitely that you are requesting an **emergency** homing.

2. Call a known station if you know the approximate locality, but call for **any** station if it is an extreme emergency.

Many a pilot who could easily have been helped by the use of homing facilities has been lost because he failed to call **any** station that might have been listening and failed to make clear that he wanted an **emergency** homing. Other stations within hearing failed to take a fix on him because they thought he was only practicing.

If by any chance your VHF receiver is not working on the homing frequency, but you are able to transmit, inform the homing station accordingly, and request that a bearing be given to you on low frequency (200-400 kc), which you can pick up on your Detrola.

If your radio fails after you have received a fix, stay on your course and be on the lookout for the field.

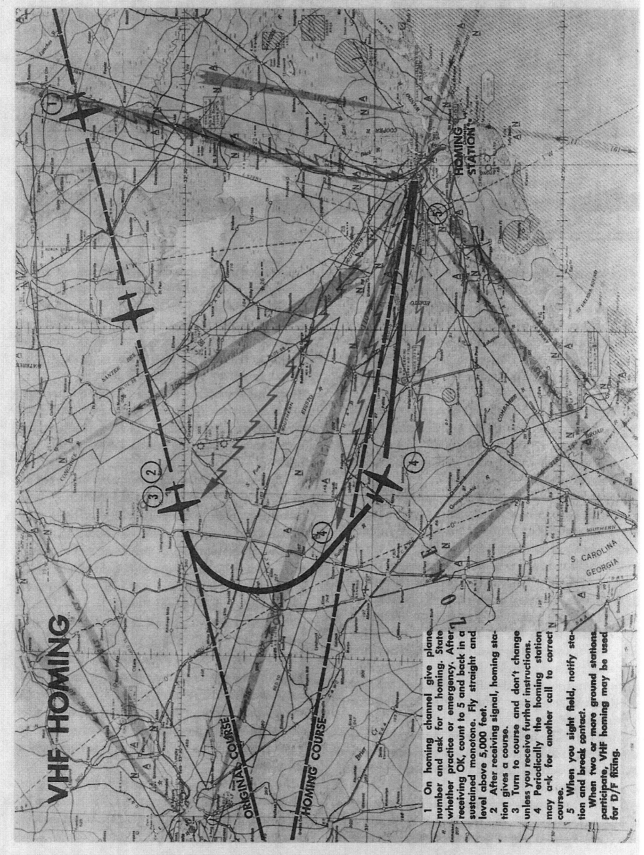

RADIO EMERGENCY PROCEDURE

The most important thing to remember if you ever have to send an emergency message is:—**speak slowly** and **don't shout.**

If the emergency is at all serious you naturally want to say a lot in a few moments. If you aren't careful, however, you're likely to talk so fast and so loud that your message is garbled.

Also, don't be afraid that the people who pick up your message will judge the seriousness of your situation by the tone of your voice. You get just as much attention if you are calm and collected as you do if you shout.

In sending an emergency message, it is a good idea to use both the channel you are already on **and** the B (American common) channel, depending upon the urgency of the situation and the amount of time you have. All homing and VHF stations in the country stand by on B channel to relay emergency messages.

To free your hands if you want to transmit only, switch the lever on the VHF control box to T, the transmitting position. In this position you don't have to hold down the transmitter button on the throttle. Be sure your throat mike is firmly against your throat, above your Adam's apple.

Then call Mayday. Be sure to give your plane number or identification. Also give your approximate location, and, if possible, the nature of your trouble.

If you are making a forced landing or going to bail out, say so. That's the best way to make sure that help reaches you promptly. Rescue parties naturally concentrate their search when they know about where you are down.

In bailing out, put your VHF control switch on T and make sure you're on the B channel. Then, by means of the carrier wave, homing stations can get a fix on your plane as it goes down, even after you have left it.

Remember: On B channel you can give an open Mayday to **all** homing stations so that anyone receiving your call can give you a homing. You aren't limited to calling any individual station.

If you are in distress and your receiver isn't working, try transmitting regardless. Your transmitter may be working and others may pick up your signals. Even if you can't carry on a 2-way conversation, if you can report yourself, you will at least get the field cleared for an emergency landing, or if you come down elsewhere, get help promptly.

Don't hesitate to call if there is any emergency.

And once you call in for a landing in an emergency and get a clearance, go straight in and land without losing any time in circling the field or making any fancy traffic pattern. Get down as soon as possible. If you have time to circle the field, why call for an emergency landing?

If you hear an emergency message being transmitted, keep off the air. Give the man who is sending it a break—don't block out his message.

RESTRICTED

OXYGEN SYSTEM

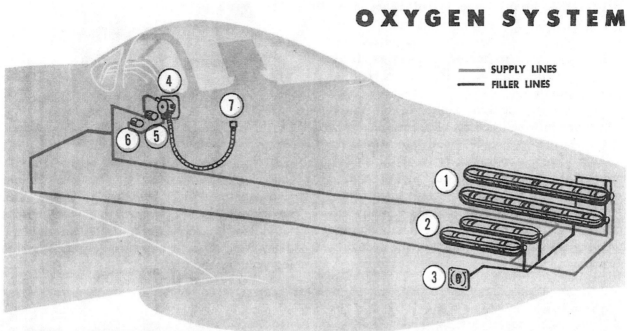

— SUPPLY LINES
— FILLER LINES

1 LOW PRESSURE OXYGEN CYLINDER TYPE F2
2 LOW PRESSURE OXYGEN CYLINDER TYPE D2
3 FILLER VALVE
4 OXYGEN REGULATOR
5 PRESSURE GAGE
6 BLINKER FLOW INDICATOR
7 OXYGEN MASK TUBE

The oxygen system in the P-51 is the same as that used in all modern army fighter planes.

It is a low-pressure, demand-type system—that is, you don't have to control the oxygen manually as you change altitudes. A regulator furnishes the right amount of oxygen required at any altitude. It does this automatically—all you have to do is inhale in your mask.

Controls and gages for the oxygen system are in the right front section of the cockpit. These include:
—the automatic mixture regulator,
—a pressure gage, and
—a blinker indicator which indicates the flow of oxygen. The blinker opens when you inhale and closes when you exhale.

You can use several different types of masks with this equipment, including the A-10, A-12, and A-14.

Notice in the illustration that there are two controls on the automatic mixture regulator. The lever on the right turns the automatic mixing device on and off. For all normal oper-

ations it should be in the NORMAL OXYGEN position. Turn it to the 100% OXYGEN position if you want pure oxygen on demand. In this position the air intake is shut off and you breathe pure oxygen on demand at any altitude.

RESTRICTED

The other knob, which is red, is an emergency control. By turning this knob to the open position you bypass the regulator and receive a **continuous** flow of pure oxygen. If the tanks are full, you get a flow of oxygen for about 8 minutes.

Your oxygen supply is carried in four tanks, which are installed just aft of the fuselage fuel tank. There are two D-2 and two F-2 tanks (which have twice the capacity of the D-2's), for a total volume of 3000 cubic inches. A filler valve, accessible through a small door in the left side of the fuselage, permits refilling the oxygen tanks without removing them from the airplane. Normal full pressure of the system is 400 psi.

If oxygen comes in contact with oil or any material containing oil, spontaneous combustion and explosion are sure to occur. Take every precaution, therefore, to keep oil, grease, and all such readily combustible materials well away from all your oxygen equipment, including your mask.

APPROXIMATE OXYGEN SUPPLY

OXYGEN CONSUMPTION DEPENDS ON SO MANY VARYING FACTORS THAT ONLY AN APPROXIMATE TIME OF SUPPLY CAN BE GIVEN. THESE TIMES ARE BASED ON A 400 PSI INITIAL PRESSURE IN THE SYSTEM

ALTITUDE	NORMAL OXYGEN	100% OXYGEN	EMERGENCY
40,000	11.4 HRS.	11.4 HRS.	12.6 MIN.
35,000	8.1	8.1	12.6
30,000	6.0	6.0	12.6
25,000	6.0	4.9	12.6
20,000	7.1	3.3	9.0
15,000	8.1	2.7	9.0
10,000	10.2	2.1	9.0

OXYGEN AND HIGH-ALTITUDE FLYING

In high-altitude flying, never underestimate the importance of the correct use of oxygen. Have a mask that is properly fitted and inspected and check it frequently for leaks. Don't be satisfied that your mask is okay simply because it was all right when you first got it.

When you're going on a high-altitude mission, check your mask thoroughly:

1. Put your mask on and check for leaks by holding the palm of your hand over the male hose fitting and inhaling gently. The mask should tend to collapse on your face.

If the mask leaks, or if there seems to be anything wrong with it, don't try to correct the trouble yourself. See your Personal Equipment Officer.

2. Insert the male fitting of the mask into the metal coupling from the regulator. Be sure that the fit is snug.

3. Inspect the rubber mask-to-regulator tubing for any damage, such as tears, holes, and kinks. Be sure that the knurled collar at the outlet end of the regulator is tight.

4. Open the emergency knob slightly and check the flow. There should be a small drop in the pressure of the system. Pressure is normally about 400 psi when the system is full (300 psi if the tanks are shatterable type).

5. Turn the emergency knob off.

6. Turn the automatic mixture regulator to the 100% OXYGEN position. Make sure that, on inhaling, the regulator diaphragm goes in and that you get pure oxygen. Check the blinker indicator.

7. Turn the automatic mixture regulator to the NORMAL OXYGEN position.

When you're using your oxygen mask, be sure that the hose is firmly connected and that it doesn't become kinked or twisted. See that the clip is attached to your flying suit or parachute harness, with sufficient hose free to allow full head movement. If it's cold enough for ice to form in the mask, flex it with your fingers to break up the ice. However, ice rarely forms in a demand-type mask.

For additional information about the use of oxygen, see your PIF.

RESTRICTED

ARMAMENT

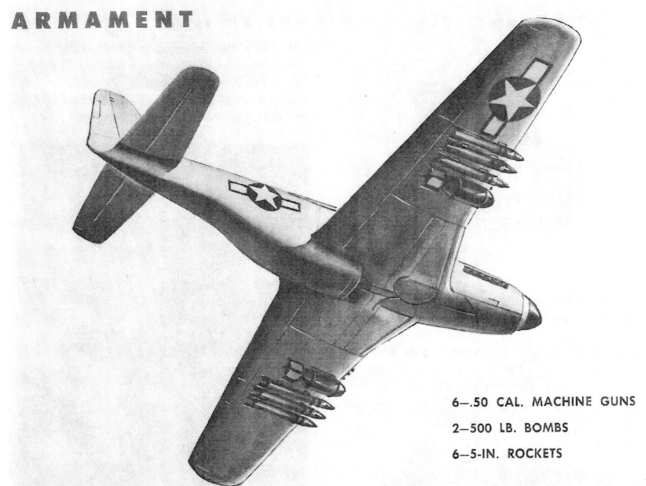

6—.50 CAL. MACHINE GUNS
2—500 LB. BOMBS
6—5-IN. ROCKETS

Gunnery Equipment

The main objective of a fighter airplane, as the name implies, is to fight—and that means gun-fight. Although its ability to carry bombs and rockets is of great importance, a fighter is primarily a flying gun platform—a means of taking firepower into the air.

Your success, then, as a fighter pilot will be measured not alone by how well you fly the Mustang, but by how well you use its guns.

Guns

The P-51D carries six free-firing .50 caliber machine guns, three in each wing. These guns are manually charged on the ground, and fire simultaneously when you press the trigger switch on the front of the control stick grip.

The maximum ammunition capacity is 400 rounds for each of the inboard guns, and 270 rounds for the center and outboard guns. This gives you a total ammunition load of 1880 rounds. The guns are adjustable on the ground, so that they can be harmonized to different patterns for various tactical situations. Usually they are aligned to converge at a range of from 250 to 300 yards.

An alternate installation is possible to meet situations where duration of fire rather than weight of firepower is of paramount importance. The center guns are removed; this allows 500 rounds of ammunition to be carried for each of the outboard guns. Your battery is thereby reduced to four guns, but the total ammunition load remains about the same.

This manual isn't intended to give you instructions in gunnery, but here's a tip—before you take off on a gunnery mission, be sure your guns are correctly loaded and charged, and that you know how fully loaded they are. There's no way of counting the number of rounds once you're in the air.

40   RESTRICTED

Gun Camera

A gun camera is mounted in the leading edge of the left wing, and is accessible for loading and adjustment from the left wheel well. A small door covers the camera aperture in the wing; this door remains open during flight, but is closed by a mechanical linkage when the landing gear is extended, thus protecting the lens from blown sand or pebbles when the airplane is on the ground.

Guns and camera are controlled by a three-position switch on the front switch panel; this same switch also turns on the lamp in the optical sight. With the switch flicked up to GUNS, CAMERA & SIGHT, the guns fire and the camera operates when you press the trigger on the stick. When the stick is down to CAMERA & SIGHT you'll take pictures by pressing the trigger, but the guns will not fire. Middle position of the switch is OFF; be sure to keep it there during takeoffs, landings, and all ground operations.

The guns and camera are heated electrically so that their operation is not affected by the extreme cold encountered at high altitude. Gun heaters are controlled by a switch on the right switch panel. The camera heater is built into the camera, and works automatically whenever the camera switch is turned on.

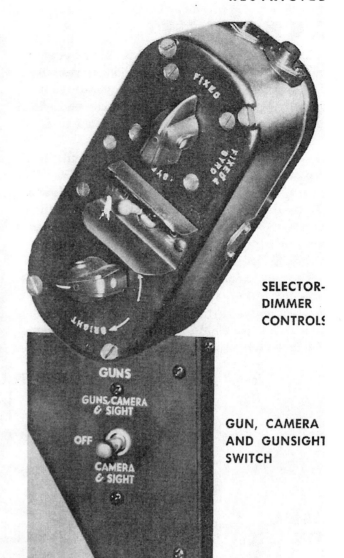

SELECTOR-DIMMER CONTROLS

GUN, CAMERA AND GUNSIGHT SWITCH

Be thoroughly familiar with your gunnery equipment and with it's limitations. The better you know your equipment, the better you know what to expect of it, in training and, most important of all, in combat.

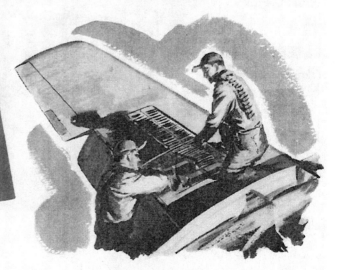

RESTRICTED

GUNSIGHT

The P-51D carries a K-14 or K-14A gunsight, mounted on the instrument hood centerline. This sight contains both fixed and gyro-actuated optical systems, and computes the correct lead angle for targets at ranges of from 200 to 800 yards.

The fixed optical system projects on the reflector glass a cross, surrounded by a 70-mil ring which can be blanked out by pulling down the masking lever on the left side of the sight. The gyro system projects a variable-diameter circle of six diamond-shaped pips surrounding a central dot.

The images are selected and their brilliance adjusted from a selector-dimmer control panel located under the left side of the instrument hood. This panel also contains a two-position toggle switch controlling the gyro mechanism. Keep this switch ON at all times—in other words, see that the gyro is operating during takeoff, flight, landing, and ground operations. This will prevent the gyro from tumbling in its gimbals and possibly damaging the gunsight.

You adjust the sight for the size of the target by means of the span scale on the front of the sight. After that, range is set into the computing mechanism by rotating the throttle grip until the diameter of the gyro image coincides with the span of the target. Targets must be tracked for at least one second before the sight will compute effectively.

The K-14 sight differs from the K-14A only in that the latter has range lines on the fixed reticle; these lines are used for aiming rockets.

Earlier airplanes of the P-51D series are equipped with the N-9 gunsight. The rheostat for this conventional optical sight is on the front switch panel.

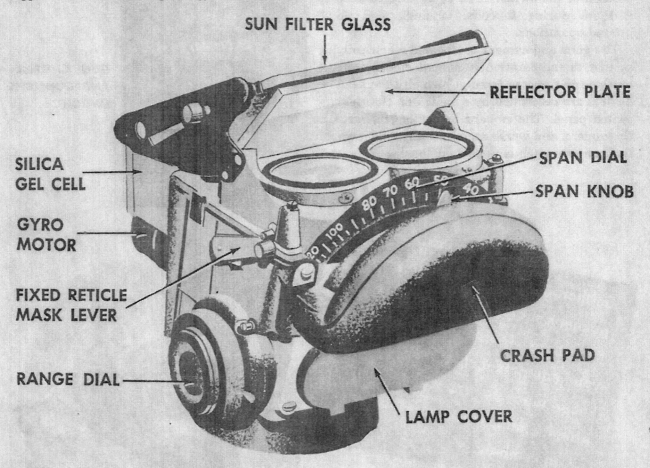

K-14A COMPUTING SIGHT

RESTRICTED

EFFECTS OF GUNNERY ON FLIGHT

The weight of ammunition carried makes no appreciable difference in the handling of the airplane, but you will notice the effect of firing the guns. With all six guns firing, you feel a slight braking action, although the actual decrease in speed is insignificant.

A one-gun jam will make the airplane yaw slightly in the direction of the wing that has all guns firing. Two- or three-gun jams, all in the same wing, will make the skid more noticeable, especially if you hold a long burst. You can still use the other guns effectively—the trick is to fire short bursts.

Don't attempt to fire long bursts and compensate for the skid by using opposite rudder. Accurate compensation is impossible and you'll just be wasting ammunition.

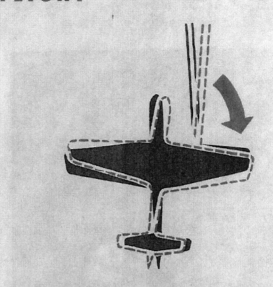

PILOT FEELS SKID TO LEFT IF LEFT GUNS JAM

ROCKETS

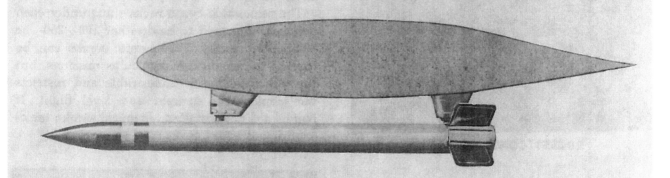

The comparatively recent development of high-velocity aircraft rockets has greatly increased the destructive capacity of fighter airplanes. Self-propelling, the missiles have no recoil and consequently can be launched from a light airplane without danger to its structure.

Later airplanes of the P-51D series are equipped to carry ten 5-inch, zero rail rockets, five under each wing. Each rocket is supported at nose and tail by a pair of launchers which are attached to the wing structure. The forward launcher contains an arming solenoid and supports the rocket by means of a forward-opening slot which engages a lug on the rocket. A safety-wired latch on the aft launcher restrains the rocket from slipping forward and falling off. When the rocket is ignited, its forward thrust shears the safety wire, allowing it to shoot forward from the launchers.

Four of the rockets are installed close to the bomb racks. Consequently, when bombs or droppable tanks are attached to the racks, only six rockets can be carried, three on each wing.

Rocket control switches are located on the front switch panel. The rockets are fired by pressing the button on top of the control stick grip. Setting the rocket release control to SINGLE or AUTO allows you to fire the rockets either one at a time or in train.

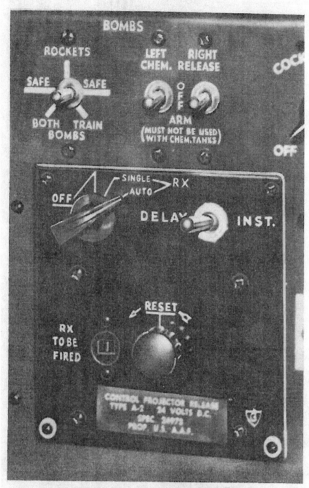

ROCKET CONTROL PANEL

Your procedure for firing rockets is as follows:

1. Turn the ROCKET TO BE FIRED dial to 1.
2. Flip the bomb-rocket selector switch to ROCKETS.
3. Move the arming switch to INST. or DELAY, as desired.
4. To fire the rockets one at a time, turn the rocket release control to SINGLE and press the button on the stick, once for each rocket.
5. To fire all the rockets in train, turn the release control to AUTO, and hold the button down. All ten rockets will be released within about one second.

Rockets cannot be jettisoned, nor can they be released in a safe condition, since the base fuse will always detonate the rocket after impact regardless of whether or not the instantaneous nose fuse is armed. Therefore, if an emergency situation (an anticipated belly landing, for example) makes it desirable for you to get rid of the rockets, use good judgment in doing so. Fire them into terrain where the resultant explosions will not endanger human lives.

BOMBING EQUIPMENT

One of the features that makes the P-51 such an outstanding airplane is its adaptability as a fighter-bomber. Though not originally designed for this purpose, its surprising load carrying capacity, ultra high speed, and excellent diving ability made it a natural for bombing assignments. As the war developed, the P-51 proved to be a better dive and skip bomber than most other airplanes, including many of those especially designed for the purpose.

The removable bomb racks slung under each wing are designed to hold either 100-, 250-, or 500-pound bombs. 1000-pound bombs can be carried to accomplish particular missions, but the extra weight is undesirable and restricts the airplane to straight and level flight. If bombs are not installed, chemical smoke tanks or droppable fuel tanks may be carried.

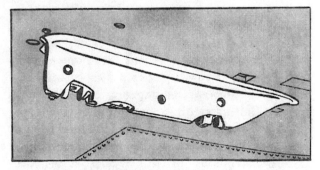

BOMB RACK

Under normal circumstances you will drop the bombs by means of the electrical release system. The bomb circuits are controlled from the front switch panel, and the release button is on top of the stick. To operate them you will:

1. Turn the bomb-rocket selector to either BOTH or TRAIN. With the switch at BOTH, the two bombs will be dropped simultaneously when the release button is pressed. At TRAIN, pressing the button will release the left bomb only; pressing it a second time releases the right bomb.

2. Flip the arming switches to ARM, unless you're jettisoning bombs over friendly territory, in which case leave the switches at OFF.

3. Press the release button on top of the control stick grip.

You can safely drop bombs when the airplane is in any flight attitude from a 30° climb to a 90° dive. But if you're sideslipping in a vertical or near-vertical dive, don't release bombs; one of them might fall into the prop and that wouldn't be good.

In case of emergency, either or both bombs can be released mechanically by pulling the salvo handles, located just below the throttle quadrant. Bombs can be salvoed in armed or safe condition, except in the event of electrical system failure, in which case the bombs cannot be armed.

Chemical tanks or droppable fuel tanks are released like bombs, either electrically or mechanically. Notice that you have to use the

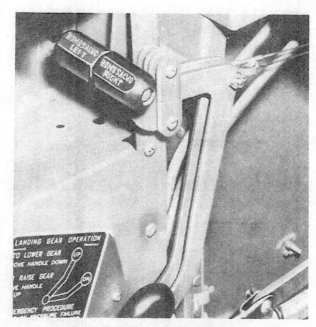

BOMB SALVO RELEASE

mechanical release if, for any reason, you want to drop the right hand bomb or tank before releasing the left.

CHEMICAL TANKS

You operate smoke or spray tanks by holding the left or right (or both) bomb arming switches in the CHEM. RELEASE position. Return the switch to OFF when you see that smoke or spray is being ejected; the action will continue automatically until the tank is empty.

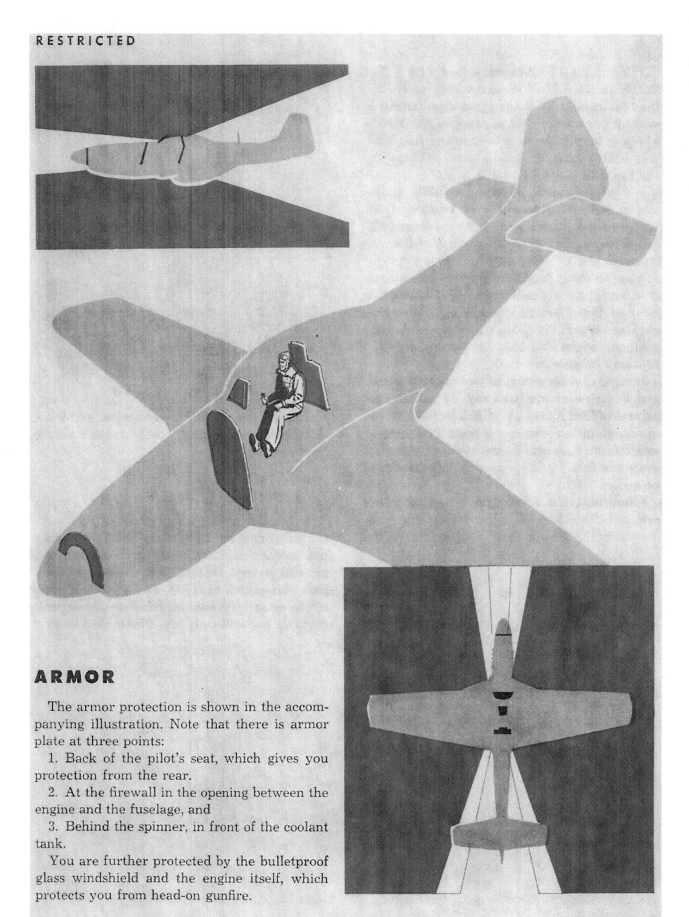

ARMOR

The armor protection is shown in the accompanying illustration. Note that there is armor plate at three points:

1. Back of the pilot's seat, which gives you protection from the rear.
2. At the firewall in the opening between the engine and the fuselage, and
3. Behind the spinner, in front of the coolant tank.

You are further protected by the bulletproof glass windshield and the engine itself, which protects you from head-on gunfire.

SIGNALING EQUIPMENT

The signaling equipment on a P-51 for use in an emergency, in the event of radio failure, or in combat zones when it is necessary to maintain radio silence, includes the following:

FLARE GUN

CAUTION: Insert the gun into the tube *before* loading it. Don't load the gun first, as there is danger of accidentally discharging the flare into the cockpit.

The flare gun, otherwise known as a pyrotechnic recognition signal pistol, is stowed in a canvas holster which is to the left and in back of the pilot's seat. There is a tube opening in the left side of the cockpit which enables you to fire the flare gun from inside the plane. The position of the tube is such that the signal is discharged in a rear upward direction.

Cartridges for the flare gun are stowed in a bag under the holster.

The flare gun can be used to indicate distress when coming in for an emergency landing, or as a recognition signal in cooperation with ground troops.

If you load the gun and then decide not to use it, always remove the flare cartridge, since there's no safety on the breech mechanism. Never leave a loaded gun in the mount, or stowed in the holster. There's no percentage in booby-trapping your ground crew.

RECOGNITION LIGHTS

Three colored recognition lights (red, green, and amber) are located on the underside of the right wing, near the tip. By means of three-position toggle switches on the electrical switch panel at your right, these lights can be used in any combination. You can burn them steadily, or flash code signals with them.

When these switches are in the down position, the lights burn steadily. When in the center position, they are off.

When in the up position, you can operate the lights intermittently, in code signals, by means of the key on top of the small box just above the switches.

DO NOT OPERATE FOR OVER 10 SECONDS ON GROUND, OR HEAT WILL MELT PLASTIC LENS

MISCELLANEOUS EQUIPMENT

TIE-DOWN KIT

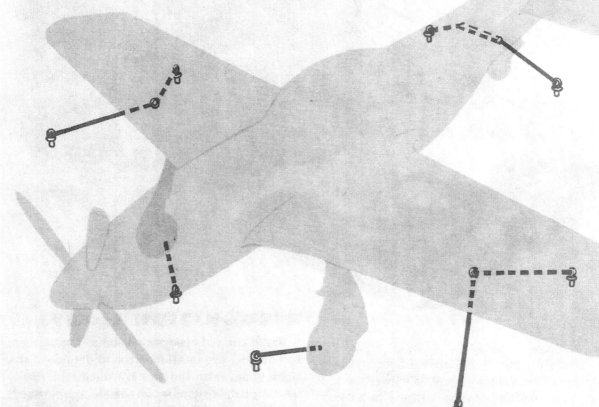

The tie-down or mooring kit is stowed in the right side of the fuselage, above the fuselage fuel tank.

The kit includes several sections of rope and standard metal tie-down stakes. A recommended method for tying down the P-51 is shown in the accompanying illustration.

DATA CASE

The airplane's data case is kept stowed in the fuselage, just aft of the oxygen cylinders.

MAP CASE AND DROP MESSAGE BAG

The map case is located just to the left of your seat. Attached to it is a fiber holder containing a drop message bag.

PILOT'S RELIEF TUBE

The relief tube is stowed on a bracket on the floor of the cockpit at the left of your seat.

FLYING THE P-51

EXTERNAL CHECK

In walking up to the plane and going around it:

1. Check the tires. See that they are properly inflated—especially that they are not too low—and not worn deeply in spots.

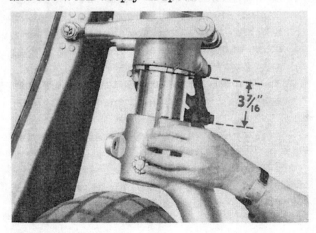

2. Check the clearances of the landing gear struts. The clearance should be about 3-7/16 inches, and equal on both struts.

3. Check the pitot tube to make sure the cover is removed.

4. Make sure the covers on the gun hatches are securely fastened.

5. Check the caps on the gas tanks, making sure they are properly closed.

In looking over the plane, check all the Dzus fasteners, especially those around the nose section. Also, be sure to check the screws in the fairings, especially those between the wing and the fuselage.

It's a good habit to approach the plane from the front and go around it in a clockwise direction, ending up at the left wing ready to hop into the cockpit.

CHECK BEFORE STARTING

As soon as you've climbed into the cockpit, make sure that the ignition switch is OFF and that the mixture control is at IDLE CUT-OFF. Have a couple of mechs pull the prop through at least 12 blades. After you have done this, make the following check around the cockpit, working from left to right:

1. Form 1-A—take the form 1-A from the case and check the status of the airplane. Make sure
   —that the airplane has been released for flight;
   —that it has been serviced with gas, oil, and coolant. If everything is okay, initial the Form 1-A.
2. Fuselage fuel—check the gage on top of the fuselage fuel tank.
3. Flap handle—UP.
4. Carburetor air control—forward in RAM AIR position.
5. Trim tabs—set as shown below.

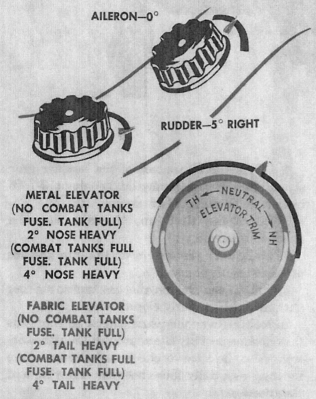

AILERON—0°

RUDDER—5° RIGHT

METAL ELEVATOR
(NO COMBAT TANKS
FUSE. TANK FULL)
2° NOSE HEAVY
(COMBAT TANKS FULL
FUSE. TANK FULL)
4° NOSE HEAVY

FABRIC ELEVATOR
(NO COMBAT TANKS
FUSE. TANK FULL)
2° TAIL HEAVY
(COMBAT TANKS FULL
FUSE. TANK FULL)
4° TAIL HEAVY

6. Landing gear handle—DOWN.
7. Left fuel gage—check gage, located on floor at your left.
8. Mixture control—IDLE CUT-OFF.
9. Propeller control—full forward to INCREASE.
10. Throttle—open to START when quadrant is so marked, otherwise open one inch.
11. Armament switches—bomb and rocket switches OFF, gun safety switch OFF, gunsight selector-dimmer switch ON.
12. Altimeter—zero or set at field elevation, as required.
13. Gyro instruments—uncage directional gyro and flight indicator.
14. Controls—Adjust your rudder pedals to comfortable position, then unlock the controls and check to see that they operate without binding. Watch the control surfaces for correct response.
15. Parking brakes—set. Don't try to hold plane with foot brakes.
16. Supercharger—AUTO.
17. Fuel shut-off valve—ON.
18. Fuel selector valve—set at the tank to which vapor return line is connected.
19. Right fuel gage—check gage, located on floor at your right.
20. Fuel booster—ON (NORMAL in earlier airplanes).
21. Ignition switch—turn to BOTH.
22. Battery and generator switches—ON.
23. Coolant and oil switches—operate manually from CLOSE to OPEN several times, and check by listening to determine whether the doors are operating. Then turn the switch to AUTOMATIC.
24. Prime and start—having completed this left to right check, you're now ready to start the engine in accordance with the procedure given below. But before you do, this may be a good time to make some of the following checks, depending on the type of mission anticipated.
   a. Before any flight, check the landing gear warning lights by pushing in on the lamp housings.
   b. If you expect to use oxygen, check the gage for a pressure of 400 psi.
   c. If night flight is anticipated, check all essential lights—instrument fluorescent lights, cockpit swivel lights, position and recognition lights, and landing lights.

STARTING PROCEDURE

After completing the Before Starting Check, proceed as follows:

1. Prime the engine three to four seconds if it is cold, one if hot.
2. Raise the starter switch cover and hold the switch at START.
3. As the engine starts, move the mixture control to RUN (AUTO RICH on some earlier airplanes). If the engine fails to take hold after several revolutions, give it one second's more prime.

Note: If the engine cuts out after starting, return the mixture control immediately to IDLE CUT-OFF.

4. Check that oil pressure goes up to at least 50 psi within 30 seconds. If it doesn't, stop the engine.
5. Idle at about 1200-1300 rpm until the oil temperature reaches 40°C and the oil pressure is steady.
6. Check the suction gage for from 3.75" to 4.25" of vacuum.
7. Check all the engine instruments. Make sure they don't exceed or fall below their limits.

After the engine is warmed up, idle at 1000 rpm or slightly less. This keeps your engine clean but not too hot.

If for any reason you expect to pull more than 40" of manifold during the engine ground run, be sure that the airplane's tail is anchored.

STARTING TIPS

1. Throttle position is important. To obtain a rich starting mixture without excessive priming, be sure your throttle is opened to the START position, or 1 inch if your throttle quadrant doesn't have the starting position designated. When the engine catches, advance the throttle to idling rpm.

With the V-1650-7 engine, the spark advance operates with the throttle. Therefore, if the throttle opening is too great, the spark will be advanced too far for starting and the prop may kick back when the engine is turned over.

2. The P-51 is easily overprimed. Use one second's prime if the engine is warm, three or four if the engine is cold. If the engine doesn't start on the first few revolutions and flames from the exhaust spread up over the cowl:

   a. Immediately cut the mixture control back into IDLE CUT-OFF.
   b. Then open the throttle gradually, and keep the starter turning the engine over.
   c. Turn off the booster pump.
   d. Turn off the starter as soon as the engine is cleared.

Do not operate the starter for over 15 seconds continuously.

If the engine is underprimed, prime it for an additional second while the engine is turning over. It should start immediately. If it doesn't, shut off the starter and repeat the process after a few moments.

STOPPING THE ENGINE

1. Put the propeller control full forward. This makes the engine easier to start next time.
2. Idle at 1500 rpm.
3. Turn the booster pump off.
4. Move the mixture control to IDLE CUT-OFF, opening the throttle as the rpm drops below 700 rpm. (Don't open the throttle above 700 rpm as any sudden opening of the throttle at this point discharges fuel into the carburetor and this causes after-firing—the engine sputters and attempts to fire again.)
5. Turn the ignition switch OFF.
6. Turn off all electrical switches. Don't forget the battery switch.
7. Lock the controls, and move the carburetor air lever to UNRAMMED FILTERED AIR.

If you are going to use the parking brakes give them plenty of time to cool or they may freeze in place. Unless you have to, better not use the parking brakes at all except in tying down the airplane for the night.

Caution: Before leaving the cockpit, be sure all the switches are in the OFF position.

TABLE OF MANIFOLD PRESSURE AND RPM LIMITS FOR FLIGHT

	TAKEOFF MAXIMUM	WAR EMERGENCY	MILITARY POWER	MAXIMUM CONTINUOUS	MAXIMUM CRUISE
MANIFOLD PRESSURE	61"	67"	61"	46"	42"
RPM	3000	3000	3000	2700	2400

TABLE OF ENGINE INSTRUMENT LIMITS

	COOLANT TEMPERATURE	OIL TEMPERATURE	OIL PRESSURE	FUEL PRESSURE
MINIMUM			50 PSI	14 PSI
DESIRED	100°-110°C	70°-80°C	70-80 PSI	16-18 PSI
MAXIMUM	121°C	105°C		19 PSI

TAXIING

When the airplane is in a 3-point attitude, the nose restricts forward visibility. This means that in taxiing you must keep S-ing continually.

Taxi with the canopy open. This not only aids visibility, but keeps the cockpit cooler on the ground.

In ordinary taxiing, keep the stick in neutral or slightly aft of neutral. This locks the tailwheel and makes it steerable 6° right or left with the rudder.

To make sharp turns or to go around corners, unlock the tailwheel by pushing the stick full forward. In this position the tailwheel is full swiveling. Be careful not to start a sharp turn before unlocking the tailwheel; it tends to bind.

If you have any trouble in unlocking the tailwheel:

1. Roll the airplane straight ahead for a short distance, and push the stick forward to release it, or
2. Wiggle the rudder controls while holding forward pressure on the stick.

Throttle back when taxiing and use the brakes as little as possible. There is no point in wasting a lot of gasoline and burning up your brakes on the taxi strip.

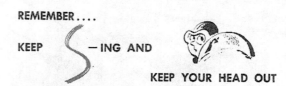

After taxiing out to the end of the taxi strip, stop at a crosswind angle so that you can look out for other airplanes and avoid dusting those behind you.

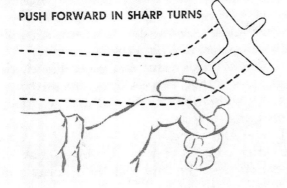

PRE-TAKEOFF CHECK

Before taxiing out onto the runway for takeoff, run up the engine and check the mags, prop, and supercharger.

1. Check the mags at 2300 rpm, and don't hold the check for longer than 15 seconds. The maximum allowable drop is 100 rpm on the right mag, and 130 rpm on the left. If either drops more than allowable, it may be that the plugs have been fouled by excessively slow engine speed in taxiing. In this case, you can often clear the plugs by the following procedure:

   a. Run the engine up to about 30° of manifold pressure. Hold this for one minute, watching the coolant and oil temperatures to see that they remain within limits.

   b. Test the mags again. If either one still drops too much, return to the line.

2. Check the prop by moving the control towards DECREASE RPM. After a drop of 300-400 rpm is indicated on the tachometer, return the prop control full forward to INCREASE RPM position. The engine should resume its former speed.

3. With the engine running at 2300 rpm, check the supercharger by holding the switch in HIGH. A drop of at least 50 rpm should be noted. Return the switch to AUTO.

LOOK AROUND BEFORE LINING-UP FOR TAKEOFF

After you've seen that the mags, prop control, and supercharger are okay, check each of the following points:

1. Coolant and oil switches—AUTOMATIC.
2. Mixture and prop controls—mixture control must be at RUN or AUTO RICH, prop control full forward to INCREASE RPM.
3. Prop and throttle friction locks—tighten these sufficiently to ensure that controls will stay put during takeoff.
4. Fuel booster—check switch at ON. If your plane has a three-position booster pump switch, flip it to EMERGENCY.
5. Hydraulic pressure—should check at 800-1100 psi with an engine speed of about 2300 rpm.
6. Canopy lever—close canopy and make sure crank lock engages. Check that red warning pins are not visible at sides of canopy.
7. Shoulder harness—lock harness and pull forward against it to be certain it's secure.
8. Engine instruments and generator—check again to see that temperatures and pressures are still within limitations, and that generator is charging.

If everything is okay you're ready for takeoff. After being cleared by the tower, make a visual check to be sure that the runway and approaches are clear.

TAKEOFF

REMEMBER: Avoid sudden bursts of power in the takeoff. Make it smooth and steady.

After you have pulled out and lined up on the runway, make sure that the steerable tailwheel is locked—it must be locked with the stick back for takeoff.

Then advance the throttle, gradually and smoothly, to 61" of manifold and 3000 rpm. Don't hoist the tail up by pushing forward on the stick until you have sufficient airspeed to give you effective rudder control.

This is important to watch in the takeoff, since the P-51, like all single-engine planes, has a tendency to turn left because of torque. If you horse the tail off the ground too quickly with the elevators, better be ready to use right rudder promptly.

Keep the airplane in a 3-point attitude until you have plenty of airspeed. In a normal takeoff, the rudder trim tab is sufficient to make the torque almost unnoticeable.

After takeoff:

1. Raise the landing gear by pulling the landing gear lever inward and up. Be sure the lever catches in the up position.
Caution: Don't brake the wheels after takeoff. Doing so may fuse the discs of brakes that are hot from extended taxiing. If this happens you'll nose up or groundloop on landing.

2. After reaching an altitude of 500 feet, throttle back to 46" of manifold at 2700 rpm.

3. Re-trim the ship for climbing attitude desired.

4. If your airplane has a three-position booster pump switch, set it at NORMAL when an altitude of 500 feet has been attained. With later planes, leave the switch ON at all times.

5. Then check over all your instruments. See that they are functioning properly and not exceeding their limitations.

In checking the instruments, make sure the ammeter indicator shows the generator is charging properly. Immediately after takeoff, the rate of charge should not exceed 100 amps, dropping back to the normal 50 amps or less after 5 minutes' operation. If it doesn't drop back to normal, turn the generator disconnect switch OFF and return to the field.

Make sure, also, that hydraulic pressure builds up to approximately 1000 psi after the landing gear has been retracted.

LANDING

On approaching the field for a landing, your best bet is to make a before-landing check before getting into the landing pattern. Anything you can do at this time in setting controls or making adjustments allows you that much more time to concentrate on the actual landing later.

As you approach the field, make the following checks:

1. Fuel—check tank gages and select fullest internal tank for landing.
2. Fuel booster—put fuel booster pump switch at ON (if your plane has a three-position switch, set it at EMERGENCY).
3. Mixture control—if it isn't already there, move mixture control to RUN or AUTO RICH.
4. Landing gear—move control to DOWN. Check indicator to see that gear is down and locked.
5. Shoulder harness—lock harness and check by leaning forward against it.
6. Prop control—forward to 2700 rpm.
7. Flaps—full down. You'll usually wait until the turn into final approach before lowering flaps.

The traffic pattern you use depends upon the local situation. Every field has its own traffic pattern, and this pattern may vary from time to time, depending upon local conditions and limitations. Therefore, no set traffic pattern is recommended or given in this manual.

There are two general rules to follow in every case, regardless of the traffic pattern. Never forget them:

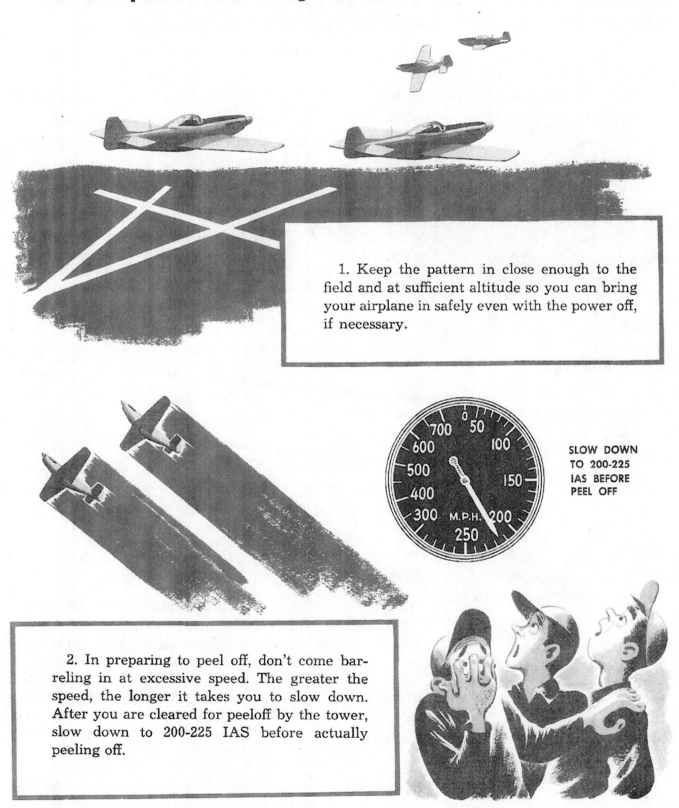

1. Keep the pattern in close enough to the field and at sufficient altitude so you can bring your airplane in safely even with the power off, if necessary.

SLOW DOWN TO 200-225 IAS BEFORE PEEL OFF

2. In preparing to peel off, don't come barreling in at excessive speed. The greater the speed, the longer it takes you to slow down. After you are cleared for peeloff by the tower, slow down to 200-225 IAS before actually peeling off.

RESTRICTED

RESTRICTED

The detailed procedure for a landing depends, of course, upon the traffic pattern. Speeds should conform to this pattern. However, here are some things to remember in any normal landing:

170 IAS BEFORE LOWERING YOUR LANDING GEAR

1. Slow down to 170 IAS before lowering your landing gear.

2. In lowering the landing gear, make sure the control handle is **DOWN and locked.** Check the landing gear indicator lights. Be sure the hydraulic pressure returns to 1000 psi. You should definitely feel each landing gear snap into position. When the landing gear comes down, the airplane gets quite nose heavy. However, you can easily adjust the trim tabs to take care of this. Don't forget that the gear takes 10 to 15 seconds to go down.

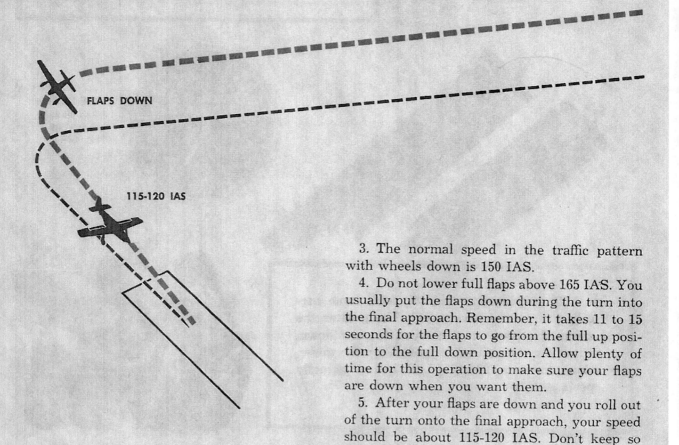

3. The normal speed in the traffic pattern with wheels down is 150 IAS.

4. Do not lower full flaps above 165 IAS. You usually put the flaps down during the turn into the final approach. Remember, it takes 11 to 15 seconds for the flaps to go from the full up position to the full down position. Allow plenty of time for this operation to make sure your flaps are down when you want them.

5. After your flaps are down and you roll out of the turn onto the final approach, your speed should be about 115-120 IAS. Don't keep so much power on that you'll be making a power

RESTRICTED

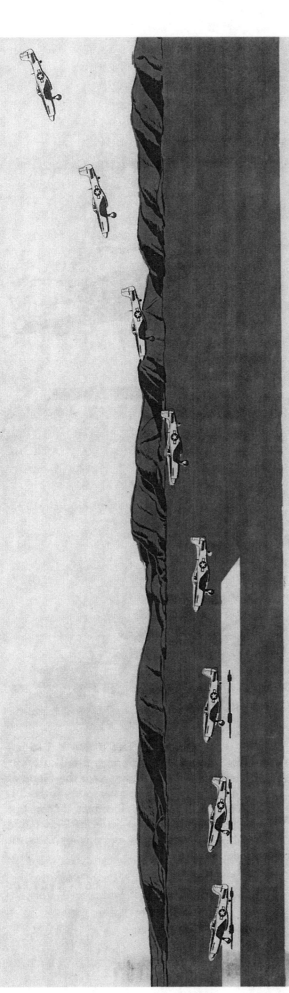

RESTRICTED

approach. However, maintain sufficient power to keep your engine clean.

6. Just before getting to the runway, break your glide, make a smooth roundout, and approach so as to land within the first third of the runway, in a 3-point attitude, as shown above.

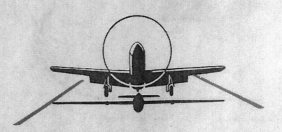

7. Hold the plane off in the 3-point attitude just barely above the runway until you lose flying speed and the plane sets down. The P-51 doesn't mush but stalls rather suddenly when you lose flying speed. So have your plane close to the runway at this point.

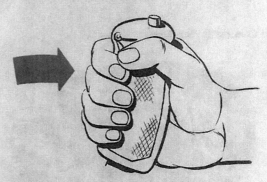

8. Since the tailwheel is locked when the stick is in neutral or to the rear, your tailwheel is automatically locked as you approach for landing. On rolling on the runway, therefore, keep the stick back until you slow down enough and are ready to turn off the runway.

9. Never attempt to push the stick forward and unlock the tailwheel in a turn. Release the tailwheel before starting the turn.

COMMON ERRORS IN LANDINGS

There are several common errors which new pilots make in landings:

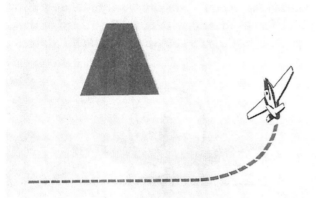

Coming in Too Low

Coming in Too Fast

Another common error is making the final turn too low and too far out. In such a situation you have to use excessive throttle to drag the plane in and have to put it into a nose-high attitude to keep out of the trees. If your engine sputters once in such an attitude, you just won't glide far.

You will come in too fast if you fail to cut your throttle soon enough, if you fail to lower your flaps in time, or if you dive the plane at the end of the runway in attempting to land it in the first third. If you're coming in too fast when you level off, you will float—and find yourself too far down the runway by the time you lose flying speed.

Coming in Too Slow

If you bring the airplane in too slowly, it tends to fall out from under you when you break your glide. You may lose your flying speed completely.

If you see you are coming in too slow, don't wait too long to use the throttle. In a high-speed airplane like the P-51, don't hesitate to use the throttle if necessary even after you have once cut it back. Your engine is there to give you power when you need it—**use it**.

Ballooning

Ballooning is a result of **overcontrolling**. You will start ballooning if you try to break a steep glide by jerking back on the stick, or if you come in fairly fast and lower your flaps too late. The airplane balloons up and you are in the embarrassing position of making a landing 10 feet off the ground.

If you ever get into this situation, ease on a little throttle and, if there is sufficient runway, land in a 3-point attitude. If there isn't sufficient runway, go around.

Catch the ballooning before you get up too high and lose flying speed. If you don't, you may drop a wing or otherwise pile up your airplane.

Bouncing

Bouncing is another result of overcontrolling. You will get into this situation if you break your glide too late, bounce on the tires, and then attempt to recover by pushing forward on the stick. You find yourself pulling back on the stick when you should be pushing forward, and pushing forward when you should be pulling back—using the right controls but at exactly the wrong time. This causes you to bounce higher and slower until you settle down permanently, the hard way.

The way to recover from such a situation is to use a little throttle at the top of the bounce. If you can't recover with the throttle and proceed with a normal landing, **go around**. Don't just pull back on the stick and sit there.

Remember, the P-51 is extremely sensitive on the elevator controls.

Forcing the Tail Down

Another common error is in attempting to force the tail down after a wheel landing while still rolling on the front wheels. If you have enough speed to keep the tail up, don't pull back on the stick, since this may cause ballooning.

If you do make a tail-high, wheel landing, the recommended procedure is to let the airplane roll until the tail settles through loss of speed.

RESTRICTED

TIPS ON LANDING

1. Check all the instruments and make all the adjustments you can while circling in the traffic pattern. This will give you more time to concentrate on the actual landing during the final approach.

2. Flare out from your approach glide into a 3-point attitude over the end of the runway. From this position you can make either a 3-point or a wheel landing.

3. Always make 3-point landings unless a wheel landing is justified by unusual weather conditions.

4. Bring the airplane down as close to the ground as possible before stalling out. The perfect position to grease it in is 1 inch off the ground.

5. Never try to land with so much speed that you have to use the stick to hold your airplane on the ground in a nose-low, tail-high attitude. Propeller clearance with the fuselage in **level** flight is only 7¾ inches.

CROSSWIND LANDINGS

The recommended procedure for crosswind landings is:

1. Drop the wing into the wind slightly to counteract the drift, and keep the plane straight with the runway.

2. Just before touching the runway, level your wings.

3. Be sure to keep the stick back after contact, so that the tailwheel will remain locked.

Make a wheel landing if the crosswind is excessive, gusty, strong, or otherwise doubtful.

Use about half flaps for any appreciable crosswind.

If you have to crab at any time, be sure to straighten out before landing. Never land in a crab as it is very hard on your landing gear.

GUSTY LANDINGS

In a gusty wind, come in slightly faster than normal.

Watch for the effect of a gust on the airplane. The gust tends to have a ballooning effect. Then, when the gust quits, the airplane is apt to drop out from under you—and you may get a wing in the ground.

Keep enough flying speed to cushion this drop.

Use about half flaps.

WET LANDINGS

The important thing about wet landings is to be especially cautious in using your brakes. Line up straight with the runway and let the airplane roll. Don't jam on the brakes, or you will probably skid into trouble.

If you come in through rain, the front of the windshield is pretty well blanked out. In that case, visibility is best out of the front panels on each side of the windshield.

MUDDY FIELD LANDINGS

The thing to remember when landing in a muddy field is to keep a definite 3-point attitude and stay off the brakes.

If you land in a tail-high attitude, your wheels will sink or drag in the mud and you'll probably nose over.

GO-AROUND PROCEDURE

Don't hesitate to go around if there is any possibility of getting into trouble while landing. It's done in the best of families. The recommended go-around procedure is:

1. Advance the throttle quickly but smoothly to a manifold pressure of 46" at 2700 rpm.
2. At the same time, counteract left torque by using right rudder and right trim tab.
3. Then trim the airplane to relieve the elevator pressure.
4. Raise the landing gear.
5. After your IAS reaches 120 mph, and you have attained an altitude of 500 feet, raise the flaps. Bring them up gradually, about 10° at a time. Watch the change of attitude as the flaps are raised.

Remember: Don't jerk or jam on the throttle. Use all controls smoothly, and pull up gradually to avoid risking a stall.

If you have rolled the elevator trim tab back for the intended landing, it may take considerable forward stick pressure to keep the nose down until you can re-trim the plane.

Most important of all in going around, continue on a **straight** course. Don't attempt any turns until your flaps are up.

FLIGHT CHARACTERISTICS

The P-51 is one of the sweetest-flying fighter planes ever built. It is very light on all controls and stable at all normal loadings.

Although light on the controls, it is not so sensitive that you would call it jerky. Light, steady pressures are all you need to execute any routine maneuver.

At various speeds in level flight or in climbing or diving, the control pressures you have to hold are slight and can be taken care of by slight adjustments on the trim tabs. However, the trim tab controls are sensitive; use them carefully. The rudder and the elevator trim change slightly as the speed or the power output of the engine changes.

The airplane is entirely normal in its flying characteristics. If you've trimmed for normal cruising speed, the airplane will become nose heavy when you raise the nose and decrease airspeed.

Under the same normal cruise conditions, when you lower the nose and increase speed, the airplane becomes tail heavy in direct proportion to the speed.

When you lower the flaps, the airplane becomes nose heavy.

When you raise the flaps, the airplane becomes tail heavy.

When you retract the landing gear, the airplane becomes tail heavy.

When you lower the landing gear, the airplane becomes nose heavy.

MAX. TAKE-OFF MANIFOLD PRESSURE— 61 IN. HG (155 CM. HG)
OPERATING RANGE 26-36 IN. HG (6604-91.44 CM. HG)

MAX. PERMISSIBLE OIL TEMPERATURE 90°C (194°F)
OPERATING OIL TEMPERATURE 70-80°C (158-176°F)
MIN. PERMISSIBLE OIL PRESSURE 50 LBS./SQ. IN.
OPERATING OIL PRESSURE RANGE 70-80 LBS./SQ. IN.

MAX. FUEL PRESSURE 19 LBS./SQ. IN.
MIN. PERMISSIBLE FUEL PRESSURE 14 LBS./SQ. IN.
OPERATING FUEL PRESSURE RANGE 16-18 LBS./SQ. IN.

MAX. TAKE-OFF RPM 3000
OPERATING RANGE 1600-2400

MAX. PERMISSIBLE INDICATED AIRSPEED 505 MPH (808 KM PH, 440 KNOTS)

CARB. AIR
DESIRABLE CARB. AIR TEMP. RANGE 15-30°C (59-86°F)
MAXIMUM 40°C (104°F)

MAX. COOLANT TEMPERATURE 121°C (250°F)
OPERATING RANGE 100-110°C (212°F TO 230°F)

Limitations for the airplane are given in the illustration above. These limitations are for all normal flying.

For your convenience, maximums and minimums for the engine are given in the airplane on a placard on the right side of the cockpit. Flight limitations for the airplane are also given on this placard.

The P-51 does not hold a sustained sideslip.

The aileron control is not sufficient to hold the airplane in a sideslipping angle. However, you can sideslip it long enough to avoid enemy fire in combat. When any sideslipping is attempted, be sure to recover completely above 200 feet.

CAUTION

In designing the later models of the P-51 and adding new equipment such as radio units and an additional gas tank, the center of gravity of the airplane has been moved back. The effect is that the amount of back pressure necessary to move the stick has been reduced. Instead of a force of 6 pounds per G of acceleration, you exert a force of only 1½ pounds, the stick forces reversing as acceleration exceeds 4 G's.

This means that you'll have to be careful in sharp pullouts and steep turns. You can easily black out—and you can also put greater loads on the airplane than its structure was designed to withstand.

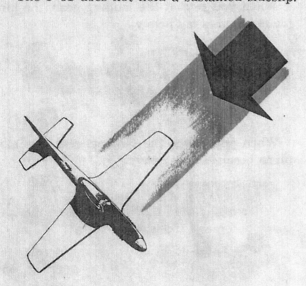

FULL FUSELAGE TANK

Be especially careful in handling the stick when the fuselage tank contains more than 25 gallons of gas. In this case the flying characteristics of the airplane change considerably—increasingly so as the amount of fuel in the tank is increased.

When you are carrying more than 40 gallons of fuel in your fuselage tank, do not attempt **any** acrobatics. The weight of this fuel shifts the center of gravity back so the airplane is unstable for anything but straight and level flight.

Be sure you are accustomed to the changed flying characteristics of the airplane before engaging in any maneuvers with a full fuselage tank. You need at least one or two hours of flying with the plane in this condition to accustom yourself to it.

REVERSIBILITY

With the fuselage tank full, the center of gravity of the airplane moves back so far that it is almost impossible to trim the airplane for hands-off level flight. Also, as soon as you enter a tight turn or attempt a pullout, the stick forces reverse.

For example, in a turn you naturally start out by holding back on the stick. But soon you find the airplane wanting to tighten up, and you have to push **forward** on the stick to prevent this.

The same thing happens in a dive. The airplane tends to pull out too sharply, and you have to change from holding back on the stick to pushing **forward** on it to keep the airplane in a proper pullout.

This is called reversibility. You'll encounter it in the P-51 only when the fuselage tank has a considerable quantity of gas in it. Be prepared in this situation. It is easily handled; just don't be surprised when it happens.

The stability of the airplane improves rapidly as you use up the gas in the fuselage tank. By the time the tank is half empty, only a slight tendency to tighten up is noticeable. It still is impossible to trim for hands-off level flight at this time, but this condition rapidly disappears as the fuel in the tank drops below the half-full level.

The P-51D's reversibility characteristics have been improved by the addition of a 20-pound bobweight to the elevator control system bellcrank. This weight reduces the amount of forward pressure you'll have to exert to overcome the reversibility tendency.

WITH EXTRA WING TANKS

When the airplane is carrying droppable fuel tanks, only normal flying attitudes are permitted. Don't try anything but normal climbing turns and let-downs when you're carrying extra wing tanks.

LOW LEVEL FLIGHT

When you're flying on the deck, trim the plane for a slightly tail-heavy condition. By doing so you'll avoid the risk of flying into the treetops if your attention is momentarily distracted from the controls.

HIGH-ALTITUDE CHARACTERISTICS

The high-altitude characteristics of the P-51 are equal to those of any other fighter plane, and in many respects are superior. With the 2-stage, 2-speed supercharger in operation, there is plenty of power up to well above 35,000 feet.

As in any airplane, the higher you go, the farther you have to move the controls to get the same results. To make a turn at 35,000 feet, for example, you have to move the controls considerably farther than to make the same turn at 10,000 feet, if your true airspeed is the same in both cases. The air up there is so thin that it takes a lot more of it to exert an equal pressure on the control surfaces.

The supercharger blower will automatically shift into high speed at between 14,500 and 19,500 feet. This change will be accompanied by a momentary power surge and increase in manifold pressure, until the manifold pressure regulator catches up.

There is no noticeable effect when the supercharger shifts back on the let-down. Therefore, below 12,000 feet notice the amber light next to the supercharger switch. If the light isn't out below that altitude, raise the cover and turn the switch to LOW.

When the supercharger is in high blower, be especially careful to handle the throttle smoothly. Any rough handling causes the engine to surge. And any surging of power above 35,000 feet greatly decreases the efficiency of the airplane and increases the effort that you have to make in controlling it.

TESTING HIGH BLOWER ON THE GROUND

HIGH-SPEED DIVING

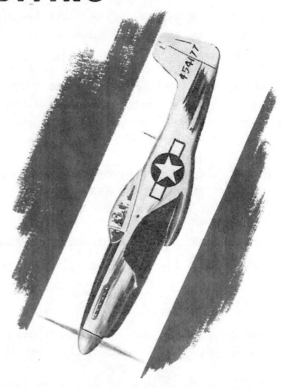

The diving characteristics of the P-51 are outstanding. Because of its clean-lined design, laminar-flow wing, exceptional aerodynamic characteristics, and small frontal area made possible by the single in-line engine, the P-51 outdives just about any airplane built.

It is capable of developing terrific speeds which makes it no toy to be played with. Yet its handling, even in high-speed dives, is not difficult if you know what you're doing.

In making a high-speed dive the most important thing is to take it easy.

Since the dive, from beginning to end, is over in a matter of seconds, you don't have much time to think things out. So know exactly what you're going to do, and then do it carefully and cautiously. Above all, don't get excited.

As the ground comes up toward you terrifically fast, allow yourself plenty of altitude for the recovery. Don't dive too close to the ground.

Note the accompanying table which shows the minimum safe altitude required for pullout from dives of various degrees. These figures are based on a constant 4G acceleration, which is about what the average pilot can withstand without blacking out.

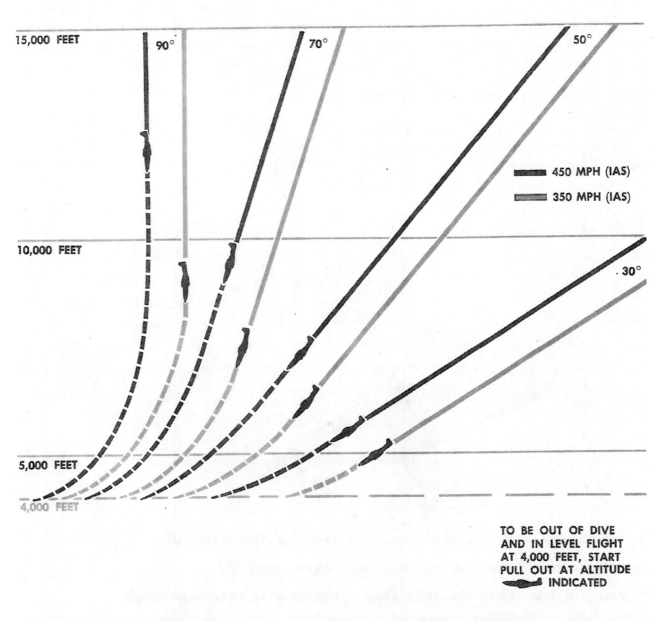

TO BE OUT OF DIVE AND IN LEVEL FLIGHT AT 4,000 FEET, START PULL OUT AT ALTITUDE INDICATED

DIVE RECOVERY PROCEDURE

The recommended procedure for recovering from a high-speed dive is:

1. Reduce the power. Don't attempt to pull out of a dive with the power on. With power on, the airplane continues to pick up speed.

2. Maintain a straight course by use of the rudder. The airplane has a tendency to yaw slightly to the right in a dive so you have to counteract this with slight use of left rudder. Don't allow the airplane to yaw, and never attempt to slow down your airplane by **deliberately** yawing it.

3. Ease the stick back. **Don't jerk the stick or otherwise overcontrol at this time.** Be sure you don't pull out abruptly.

NOTE that in this recommended dive recovery, you don't use the trim tabs. It isn't necessary to use the tabs, and since they are extremely sensitive, it is recommended that you don't use them. With the airplane trimmed for normal cruise, you can control the airplane in a high-speed dive with only the stick and rudder pedals.

In extremely high-speed dives, you can use the trim tabs intentionally, if you desire, but use them carefully.

If you use the tabs, the following procedure is recommended:

1. Trim the airplane for normal cruising.
2. About halfway through the dive, use slight elevator and rudder trim, but be careful not to trim the airplane nose heavy.
3. As the airplane continues to accelerate, it again becomes tail heavy—increasingly heavy as speed increases. However, make no further adjustment of the tabs. After having made this one adjustment, you can control the airplane easily with the stick and rudder. The ailerons become increasingly heavy as the speed of the airplane increases.

MAXIMUM ALLOWABLE IAS

The maximum safe airspeeds for the P-51 at different altitudes are given in the accompanying graph. Note that the figures given are IAS (indicated airspeed) figures

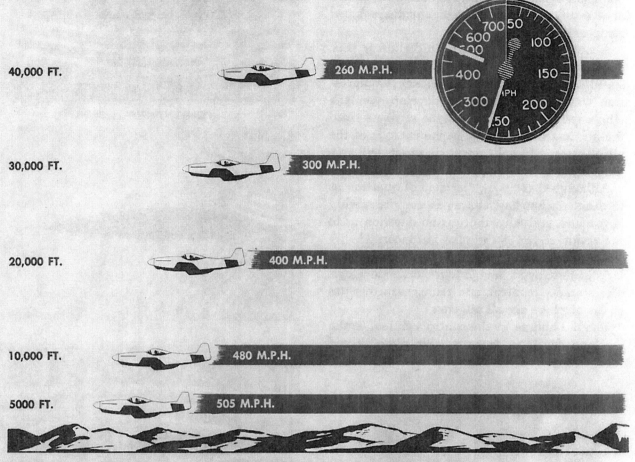

NEVER EXCEED THESE SPEEDS

If you do, you're asking for trouble

Notice that at altitudes above 5000 feet the figures are **less** than 505 IAS (the red-line figure on the airspeed indicator of the airplane). At 40,000 feet, for example, the maximum safe speed is 260 IAS.

In other words, the red-line figure for the P-51 is not a **fixed** figure but a **variable** figure—variable with altitude. The higher you go, the lower the maximum allowable IAS. This is true of all ultra-fast, high-altitude fighter planes used for high-speed diving.

The **usual** red-line speed for an airplane (the one marked on the airspeed indicator) is the speed at which the airload on the wings and other structural members reaches the maximum that these members are designed to carry. Above this speed, the wings and other structural members cannot safely carry the extreme airloads that develop.

In the case of high-speed fighter planes, however, a new factor enters the picture which makes diving unsafe at high altitudes long before the usual red-line speed is reached. This new factor is compressibility. It is the reason—and a good one—for the variable red-line speed above 5000 feet.

COMPRESSIBILITY

Since extremely high airplane speeds have been developed only in recent years, the phenomenon of compressibility is still pretty much of a mystery. Scientists and engineers know comparatively little about it.

About all that is known for certain is this: Just as soon as an airplane approaches the speed of sound, it loses its efficiency. Compression waves or shock waves develop over the wings and other surfaces of the airplane. And the air, instead of following the contour of the airfoil, seems to split apart. It shoots off at a tangent on both the upper and lower surfaces.

Although there is a great deal of question as to exactly **what** happens when compressibility is reached, and **why**, there is no question as to the **result**, so far as the pilot is concerned.

The lift characteristics of the airplane are largely destroyed, and intense drag develops. The stability, control, and trim characteristics of the airplane are all affected.

The tail buffets, or the controls stiffen, or the airplane develops uncontrollable pitching and

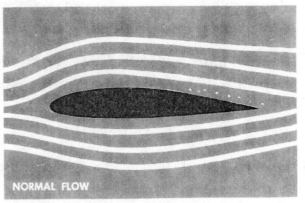
CONVENTIONAL AIRFOIL

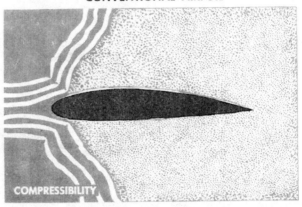

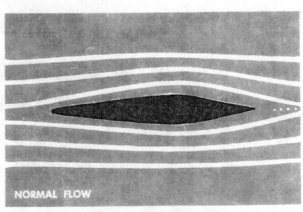
LAMINAR FLOW AIRFOIL

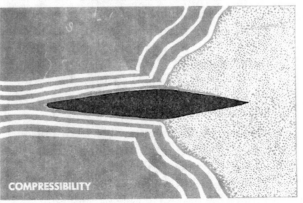

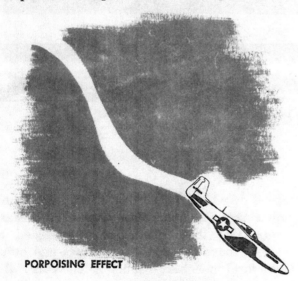
PORPOISING EFFECT

porpoising, or uncontrollable rolling and yawing, or any combination of these effects. Each type of high-speed fighter plane has its own individual compressibility characteristics.

If the speed of the airplane isn't checked and the pilot doesn't regain control of it, either the terrific vibrations of the shock waves cause structural failure or the airplane crashes while still in the compressibility dive.

In your P-51, the first effect of compressibility that you feel is a "nibbling" at the stick—the stick will occasionally jump slightly in your hand. If you don't check the airspeed, this will develop into a definite "walking" stick—the stick will "walk" back and forth and you won't be able to control it. At this stage the airplane is beginning to porpoise—that is, to pitch up and down in a violent rhythm like a porpoise. As the airplane accelerates further, the porpoising will become increasingly violent.

Once the airplane begins to porpoise, you won't be able to anticipate its porpoising movements by any counter-movements of the stick. Anything you do in this regard merely makes the situation worse. Or you may develop an aggravated case of reversibility—the control forces reverse, as they do when your fuselage tank is full and you have to push forward on the stick in a dive to keep the airplane from pulling out too abruptly.

It is possible to come out of compressibility safely if you don't go into it too far. But before discussing the recovery procedure, here are some additional facts about compressibility.

MACH NUMBER

An airplane goes into compressibility before actually reaching the speed of sound. Some airplanes go into it when they reach 65% of the speed of sound; some when they reach 70% of the speed of sound. It all depends on the design of the airplane.

The percentage figure at which any particular airplane goes into compressibility is known technically as its critical Mach number (named after the man who discovered this relationship between true airspeed and speed of sound).

The P-51 has one of the highest critical Mach numbers of any airplane now in combat. It can be dived to beyond 75% of the speed of sound before going into compressibility.

One of the most important factors to remember about compressibility is that the speed of sound varies with altitude. Note these approximate figures:

At sea level, sound travels 760 mph.
At 30,000 feet, sound travels 680 mph.

The higher you are, therefore, the sooner you approach the speed of sound. And, the higher

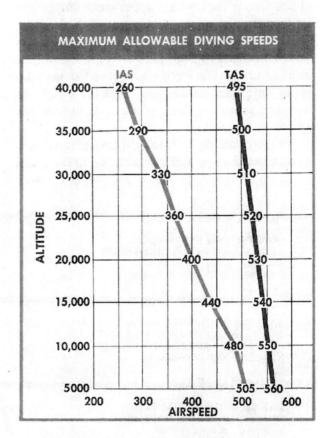

you are, the lower your safe IAS

When you get above 5,000 feet in the P-51, the maximum safe IAS is less than the 505 IAS red line of the airplane. Above that altitude, you go into compressibility before you reach the red line on your airspeed indicator. That's the reason for the variable red line speed as given in the graph.

The accompanying illustration shows the maximum allowable safe speeds in terms of TAS as well as IAS. Notice how much these two figures differ. At 35,000 feet, for example, an IAS of 290 mph means you're actually traveling 500 mph (TAS)!

Many a pilot fails to realize this great difference between IAS and TAS at high altitudes. Don't be fooled—study these figures carefully.

UNCONTROLLED DIVE

As noted earlier, it is possible to come out of compressibility safely if you don't go into it too far. The most important thing to remember about this is that while in compressibility you have virtually **no control** over your airplane. While in compressibility you can aggravate your situation, you can make it a lot worse. But outside of cutting off the power (if it isn't already off) and holding the stick as steady as possible, there's nothing you can do to **help** the situation.

All you can do is ride it through until you decelerate enough and lose altitude to the point where your speed is below the red line speed as given in the table. **This usually means an uncontrolled dive of between 8000 and 12,000 feet, depending upon circumstances.** The exact distance you drop and the length of time you are in compressibility depend to a great extent upon the angle of dive in which you **encountered** compressibility.

Only after you have lost enough speed and altitude, do you come out of compressibility and regain control of your airplane. At that point—with the airplane again completely under your control—you can **begin** to come out of your dive.

Note that last sentence carefully. You can **begin** to come out of your dive—that's after losing 8000 to 12,000 ft. If at that point you still have sufficient altitude for a controlled dive recovery, you will be okay. If not, ...?

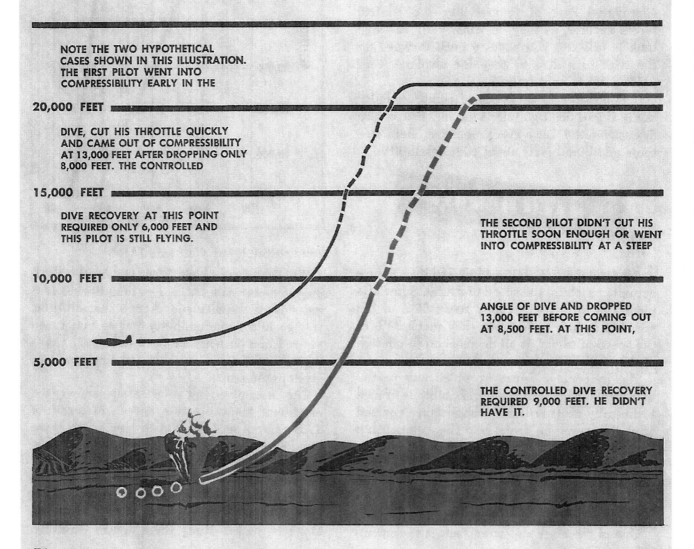

NOTE THE TWO HYPOTHETICAL CASES SHOWN IN THIS ILLUSTRATION. THE FIRST PILOT WENT INTO COMPRESSIBILITY EARLY IN THE DIVE, CUT HIS THROTTLE QUICKLY AND CAME OUT OF COMPRESSIBILITY AT 13,000 FEET AFTER DROPPING ONLY 8,000 FEET. THE CONTROLLED DIVE RECOVERY AT THIS POINT REQUIRED ONLY 6,000 FEET AND THIS PILOT IS STILL FLYING.

THE SECOND PILOT DIDN'T CUT HIS THROTTLE SOON ENOUGH OR WENT INTO COMPRESSIBILITY AT A STEEP ANGLE OF DIVE AND DROPPED 13,000 FEET BEFORE COMING OUT AT 8,500 FEET. AT THIS POINT, THE CONTROLLED DIVE RECOVERY REQUIRED 9,000 FEET. HE DIDN'T HAVE IT.

ELEVATOR MODIFICATION

Latest P-51D's and K's come from the factory with metal covered elevators and with decreased angle of incidence of the horizontal stabilizer. Existing airplanes will be modified in the field so that ultimately the changeover will affect every airplane of the P-51D and K series. Be sure you know the status of your plane because this modification changes some of the flight characteristics, at high Mach numbers, from those described on the preceding pages. Porpoising has been eliminated up to Mach number of at least .80. However, the elevator stick force characteristics are not as good. When diving a modified airplane you will find that as you get close to a Mach number of .74, less and less forward pressure on the stick is required to maintain the angle of dive. As your speed exceeds .74 Mach number, you will have to start pulling on the stick to keep the nose from dropping. This pull will continue to increase with Mach number. As an example, in a dive test performed by the Flight Section of ATSC it was found that at .775 Mach number a pull of 10 pounds was required to maintain a straight forward flight path. This stick force was an increase from 0 stick force at a Mach number of 0.746. Also, a greater additional force is required to start recovery from a dive at high Mach number than from a dive at low Mach number.

The placard Mach number limit for the modified airplane is the same as for the others—.75. So long as you don't exceed it you'll be all right, but you are sticking your neck out when you do. You won't feel serious compressibility effects if you keep your diving speed below .75 Mach number, and recovery can be made without difficulty. Exceeding that Mach number will bring on vibration of the stick, vibration of the airplane, and a wallowing motion caused by low directional stability. This means that you must start a smooth recovery. Do not wait or try to ride the dive to a lower altitude because that technique is not necessary with this airplane; smooth recovery is possible at any altitude sufficiently high.

COMPRESSIBILITY RECOVERY PROCEDURE

If you ever get into compressibility in a high-speed dive, don't get excited. Keep calm, and follow this recommended recovery procedure:

1. **Cut the power immediately.** To get out of compressibility you've got to lose airspeed, so **cut your throttle back.**
2. Release a slight amount of the forward pressure you're holding on the stick.
3. Don't allow the airplane to yaw. Never **deliberately** yaw it to slow the airplane down.
4. Hold the stick as steady as you possibly can. Don't attempt to anticipate the porpoising movement by counter-movements of the stick.
5. As the airplane slowly but steadily decelerates with power off, and you get into the lower altitudes where the speed of sound is greater, the porpoising stops and you regain complete control of the airplane.
6. Pull out of the dive in a normal recovery. **Don't pull out abruptly.** Take it as easy as altitude permits.

Notice in the above procedure that you don't use the elevator trim tab. It isn't needed.

GLIDING

You can glide the P-51 safely at any speed down to 25% above stalling speed. Under average load, this is about 125 mph IAS at any level, the speed increasing with the weight of the airplane. Although the minimum safe gliding speed increases with altitude in terms of TAS, it remains approximately the same in terms of IAS.

When the landing gear and the flaps are up, the glide is fairly flat. In this condition, however, with the nose extremely high, forward visibility is poor.

Lowering either the flaps, the landing gear, or both, reduces slightly the minimum safe gliding speed, greatly steepens the gliding angle, and increases the rate of descent.

STALLS

A stall in the P-51 is comparatively mild. The airplane does not whip at the stall, but rolls rather slowly and has very little tendency to drop into a spin. When a complete stall is reached, a wing drops. After that, if you continue to pull back on the stick, the airplane falls off into a steep spiral.

When you release the stick and rudder, the nose drops sharply and the airplane recovers from the stall almost instantly.

You'll generally be warned of an approaching stall by a buffeting at the elevators. In a power-off stall the buffeting is slight, becoming noticeable at 3 or 4 mph above stalling speed. Violence of the elevator buffet increases with the speed of the stall.

The speeds at which stalling occurs vary widely, depending on the gross weight and the external loading of the airplane. Lowering the flaps and landing gear, of course, reduces stalling speed considerably.

A **power** stall either with wheels and flaps up or with wheels and flaps down is much more violent than a power-off stall.

Notice that while in a stalling attitude the rudder remains sensitive well after the ailerons have lost their efficiency. You can see, therefore, why a sudden application of power in making a landing will aggravate a wing-low condition.

Recovery from any stall is entirely normal. Apply opposite rudder to pick up the dropping wing and release the back pressure on the stick.

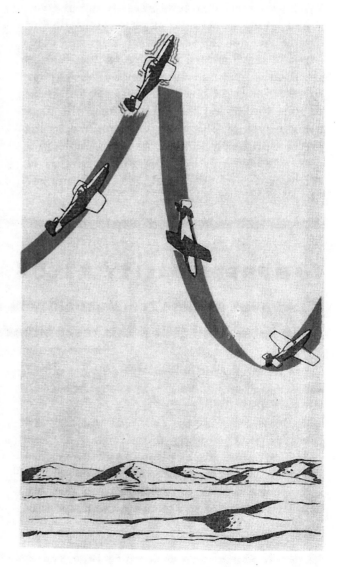

SPINS

Power-off spins in the P-51D are safe enough if you have plenty of altitude for recovery. However, you'll find them quite uncomfortable because of heavy oscillations.

When you apply controls to start a spin, the airplane snaps ½ turn in the direction of spin as the nose drops to near vertical. After one turn, the nose rises to or above the horizon and the spin slows down. The airplane then snaps again, and the process is repeated.

Spins to the left will occasionally dampen out and become stable after about three turns, but in right spins the oscillations are continuous, neither increasing nor decreasing as the spin progresses.

Power-on spins are extremely dangerous and must never be performed intentionally under any circumstances. The nose remains at from 10 to 20 degrees above the horizon, the spin tends to tighten, and there is a rapid loss of altitude. Recovery control will have no effect on the airplane until the throttle has been completely cut back.

The spin recovery procedure recommended is the standard N.A.C.A. procedure, and is the same for both left and right spins.

N.A.C.A. SPIN RECOVERY

1. **Pull the stick back and use full rudder with the spin.**

2. **Cut the throttle.**

3. **Apply full opposite rudder to slow and stop the spin.**

4. **Move the stick quickly forward to pick up flying speed.**

As soon as you apply opposite rudder the nose drops slightly and the spin speeds up rapidly for about 1¼ turns and then stops. The rudder force at first is light but then becomes heavy for about a second or so in the first half turn. The rudder force then drops to zero as the spin stops.

During the spin you feel a slight rudder buffeting. If you attempt to recover from the dive too soon after the spin stops, you also feel rather heavy buffeting in both the elevator and the rudder. The remedy for this condition is to release some of the pressure you have applied on the stick.

If you should ever get into a **power spin, cut the throttle immediately** and follow the normal recovery procedure. Be sure to hold the controls in the recovery position until you have recovered completely. It may take up to six turns to recover from a two to five turn power spin. In this situation you may lose as much as 9000 feet of altitude.

Remember these tips on spin recovery:

1. Don't get excited.

2. Don't be impatient. Leave the controls on long enough for them to take effect.

3. Fix in your mind the altitude at which to bail out, and bail out before it is too late.

4. Never make an intentional power-**on** spin.

5. In making an intentional power-**off** spin, start it with plenty of altitude. Be sure you can recover above 10,000 feet.

IMPORTANT: If the normal recovery procedure doesn't bring you out of the spin, let the controls go.

RESTRICTED
ACROBATICS

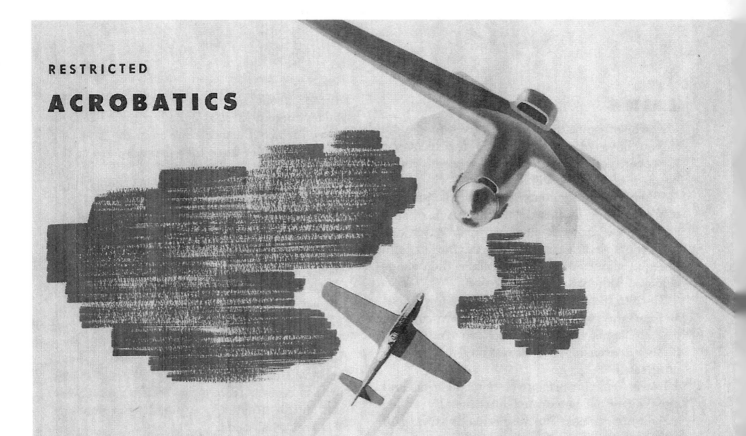

The P-51D has really exceptional acrobatic qualities; stick and rudder pressures are light and the aileron control is excellent at all speeds. Be sure of one thing before entering any acrobatic maneuver—have plenty of altitude.

You can do chandelles, wingovers, slow rolls, loops, Immelmans, and split-S turns with ease. However, remember that you must limit inverted flying to 10 seconds because of loss of oil pressure and failure of the scavenger pump to operate in inverted position.

In a loop you have to pull the airplane over the top, as the nose won't want to fall through by itself. If you don't fly the airplane on over the top of the loop, it has a tendency to climb on its back.

The aerodynamic characteristics of the P-51D are such that snap rolls cannot be satisfactorily performed. This has been proved by a long series of test flights. So don't try any snap rolls in an attempt to show that you're the guy who can do them. You'll invariably wind up in a power spin—and that's bad.

Caution: Acrobatics must not be attempted unless the fuselage tank contains less than 40 gallons of fuel.

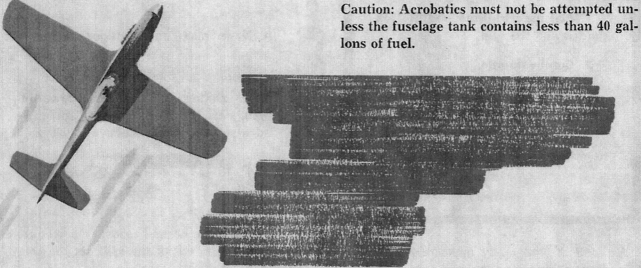

EMERGENCY PROCEDURES

FORCED LANDINGS ON TAKEOFF

If your engine fails on takeoff, immediately nose the airplane down to retain airspeed. If you have sufficient runway, simply make a nor-

mal 3-point landing straight ahead. If you don't have sufficient runway, make a belly landing.

One of the most important things to remember if your engine fails on takeoff, is to land straight ahead—or only slightly to the right or left depending on obstructions. **Never attempt to turn back into the field.** There is only a slim chance that you can make it. Steep turns near the ground are hazardous even with power on; with a dead engine they are suicidal.

In making a forced landing on takeoff when the runway is behind you, nose the plane down and maintain a glide of about 110 mph. If you are carrying droppable fuel tanks or bombs, maintain a glide of about 120 mph and salvo the auxiliary load immediately.

Duck your head and jettison the canopy. Move the landing gear control to UP. The gear may not have time to retract, but once it is started it will collapse on contact with the ground.

Use full flaps, and cut the ignition, fuel, and battery switches before contact.

FORCED LANDING OVER DOUBTFUL TERRAIN

If you have to make a forced landing and the terrain is doubtful, don't hesitate to make a belly landing. Forced landings with wheels down should be made only when you're absolutely certain that such a procedure will be safe.

BELLY LANDING PROCEDURE

If you have to make a belly landing, it is best to make the landing on a hard surface. On soft or loose ground the airscoop tends to dig in, not only stopping the plane suddenly but doing more damage to your plane than if you land on a hard surface.

The belly landing procedure is as follows:

1. Keep the wheels up.
2. Jettison tanks or bombs, if you're carrying any.
3. Lower the seat, duck your head, and jettison the canopy.
4. Make sure your shoulder harness and safety belt are locked.

5. Use about 30° of flaps until just before landing. Then, when you're sure you have your landing area well in hand, use full flaps.
6. Maintain a speed of about 120-130 mph until actually landing.

7. Approach in a 3-point attitude to slow the airplane.
8. Cut the switches just before impact.

9. As soon as the airplane stops, get out and get away from it quickly.

Unless you're close to a farmhouse or town where you know you can call for help, don't wander away from your airplane. Stay near it, especially in swampy ground where you won't be able to travel far anyway. It is easier for searching parties to find your airplane than to find you alone. Also, you may want to use parts of your airplane for signaling. For example, you may want to use oil or gasoline for building a signal fire, or to use bright reflecting parts of the plane for attracting attention.

FORCED LANDING OVER AN AIRFIELD

If your engine fails directly over an established airfield, your problem is simply one of making a routine precision landing.

If you are not directly over a field but are within gliding distance of it, don't lower your gear until you are sure you can make it in, as the glide is extremely steep with wheels down.

FORCED LANDING AT NIGHT

If you ever have to make a forced landing at night, better bail out unless conditions are exceptional.

Don't attempt to bring your airplane in at night, even in a belly landing, unless you have radio contact, are right over a known airport, and feel positive that the plane is in condition to be safely landed.

If you run into serious trouble at night and are not right over an airport—bail out.

ENGINE OVERHEATING

If your engine overheats in flight, the trouble is probably caused by one of the following:

1. You've been climbing the airplane at high power and below recommended airspeed. In other words, you aren't getting a great enough blast of air through the airscoop. To remedy this difficulty, all you have to do is level out for awhile—increase airspeed but reduce power.

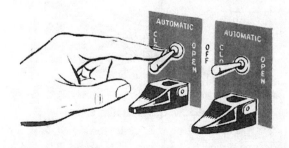

2. The automatic shutter controls are not functioning properly. In this case, operate the shutters manually by means of the toggle switch control, and watch the instruments to see if the condition has been remedied.

3. The oil supply is depleted. You discover this situation in checking the oil pressure. The engine continues to overheat even after the shutters are opened all the way. There isn't much you can do in this situation except keep the rpm and power settings at the minimum, and land as soon as possible.

4. The coolant supply is depleted. Here again, the engine continues to overheat even after the shutters are opened all the way. There isn't much you can do in this situation, either, except keep rpm and power settings at the minimum, and land as soon as possible. In most cases you won't have more than 10 minutes before the engine freezes.

5. You've been exceeding the operational limits of the engine. Make sure that the carburetor air control is at RAM AIR, depending upon the type of equipment. Then check the mixture control to see that it is in RUN or AUTO RICH.

RUNAWAY PROPELLERS

Failure of the propeller governor is quite rare, and the chances are that you will never encounter it. When it does happen the prop runs away, that is, the blades go to full low pitch, resulting in engine speeds as high as 3600 rpm or more. Obviously, this speed must be reduced immediately or the engine will be totally ruined, necessitating a forced landing or a bailout.

If you're ever confronted with a runaway prop, the following procedure is in order.

1. Pull the throttle back to obtain 3240 rpm, the maximum allowable diving overspeed of the engine.

2. Raise the nose of the airplane to lose speed, and if you're flying very high, return gradually to a moderate altitude. Keep your IAS at about 140 mph.

3. When you reach a landing field, lower the gear and make a normal landing.

BRAKE FAILURE

Remember that the brake system is not operated by the hydraulic system of the airplane and that each brake is operated by its own individual pressure cylinder, which is activated by using the brake pedals. It is extremely unlikely, therefore, that both brakes will fail at the same time. When one brake fails it is almost always possible to use the other in stopping the airplane.

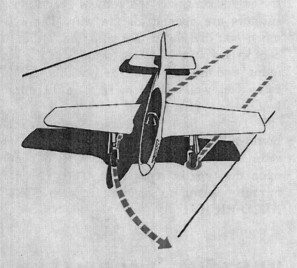

If one brake goes out while taxiing, use the other (good) brake, and also the lockable tailwheel. Immediately chop the throttle and cut the switch. If you're going too fast to stop the airplane in this manner, lock the good brake, and groundloop until the airplane stops.

If a brake goes out while checking the magnetos, immediately cut the throttle back and hold the plane in a groundloop with the good brake.

If, in coming in for a landing, you know that your brakes are shot—or even if you suspect such a condition—approach the field and land as slow as safety permits. Use full flaps and use your best technique in making a 3-point landing. Stop your engine completely by cutting the mixture control as soon as your plane is on the ground. The dead prop creates additional braking action to help make your landing as short as possible.

If your brakes are locked, never attempt a wheel-type (tail high) landing. If you do, you will either hit the prop or nose over altogether.

HYDRAULIC SYSTEM FAILURE

If your hydraulic system ever fails, remember that you can lower the landing gear by pulling the emergency knob. The procedure is simple:

LANDING GEAR DOWN

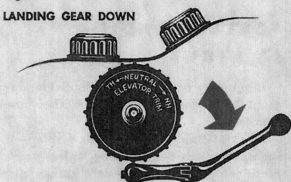

1. Put the landing gear control handle in the DOWN position. This releases the mechanical locks which hold the gear in place.

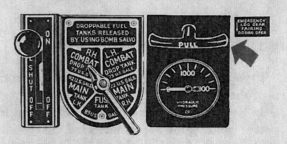

2. Pull the red emergency knob. This releases the hydraulic pressure in the lines and allows the gear to drop of its own weight.

ROCK IT TO LOCK IT

It is possible that the gear may not fall with sufficient force to lock itself in place. Therefore, **while still pulling out on the red emergency**

knob, rock the airplane until you feel the gear catch in the locked position.

The tailwheel usually locks without any difficulty. If it doesn't, speed up the airplane to force the partially extended wheel into position by means of greater air pressure on it. Or dive the airplane a short distance and then pull out with enough acceleration to force down the tailwheel.

ELECTRICAL SYSTEM FAILURE

The airplane's electrical system circuits are protected by circuit breaker switches on the right hand panel. These switches are controlled by a single bump plate hinged across them, enabling you to re-set all the buttons at one time, and doing away with the necessity of hunting for the right switch.

If you have an overload on any circuit, the breaker for that circuit will pop out. To re-set, wait a few seconds for the switch to cool, and then give the bump plate a firm push. If the switch pops out again immediately, allow a bit more cooling time and try again. Should repeated efforts fail to restore the switch to its original position, there's nothing more you can do. The trouble is probably a short circuit that cannot be repaired in flight.

If the ammeter shows that something has gone wrong with the electrical system and the battery is overcharging, cut the generator disconnect switch OFF. Be careful that you don't overcharge the battery.

If you ever have to shut off the generator, use your radio sparingly, as the radio quickly drains your battery.

If the ammeter shows that the battery is undercharging, check the generator disconnect switch to make sure that it hasn't been turned OFF accidentally. If it's still ON and the battery is not charging properly, use your radio only when necessary. Make the best use of whatever battery power is still available.

Remember: If the electrical system goes completely dead, the ignition system continues to operate on the magnetos. However, you won't be able to control the oil and coolant scoops, since they operate electrically.

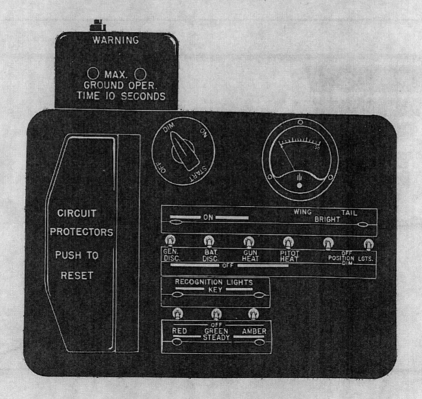

TIRE FAILURE

If, in landing, you know that a tire is low or blown out, make a 3-point landing. Don't use the brakes until necessary, then use the opposite brake—but only slightly—and enough opposite rudder to keep the airplane straight.

Land on the left side of the runway if the right tire is flat and on the right side if the left tire is flat. Then, if you swerve in the direction of the flat, you'll still be on the runway.

If a tire is completely lost, don't attempt to come in on the rim. Make a belly landing.

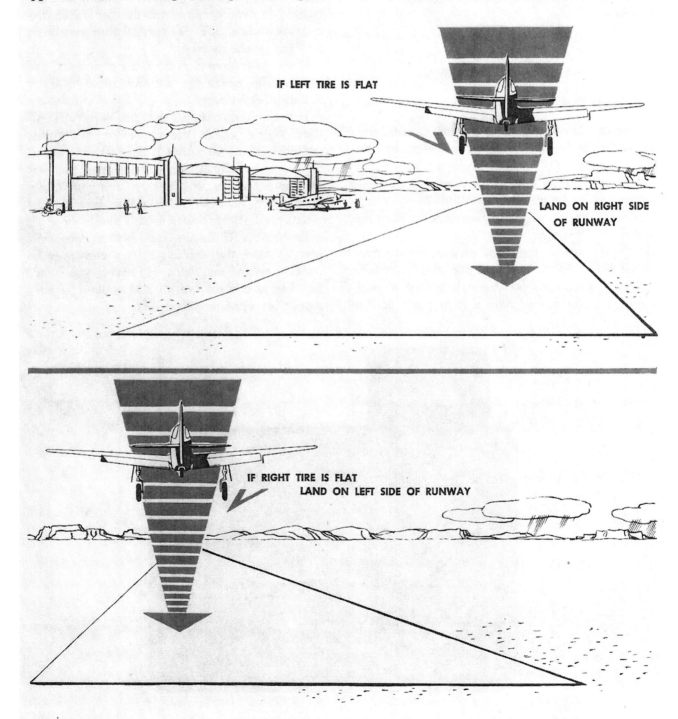

IF LEFT TIRE IS FLAT
LAND ON RIGHT SIDE OF RUNWAY

IF RIGHT TIRE IS FLAT
LAND ON LEFT SIDE OF RUNWAY

FIRE

The most important thing to remember in event of a fire is to keep the canopy entirely closed as long as you stay in the cockpit. As soon as you open the canopy you make a chimney out of the cockpit and draw the fire into it, usually through the holes in the floor just ahead of the rudder controls.

For the same reason, don't lower the gear. Opening the gear wells is likely to blast the fire into the cockpit.

If fire breaks out and you don't think it's necessary to abandon the airplane—at least right away—protect yourself by covering all the exposed parts of your body. Put your goggles down over your eyes, for example, roll down your sleeves, and otherwise protect yourself from possible flash burns.

If you decide to bail out, follow the normal bailout procedure up to the point where you release the canopy. When you do release the canopy, leave the airplane at the same instant that the canopy does. **Do not release the canopy until you're all ready to go out with it.**

Don't release the canopy until after you have unlocked the safety harness, trimmed the airplane, and are crouched with your feet in the seat ready to spring out. Then pull the canopy emergency release handle and lunge upward to the right, pushing the canopy off with your head.

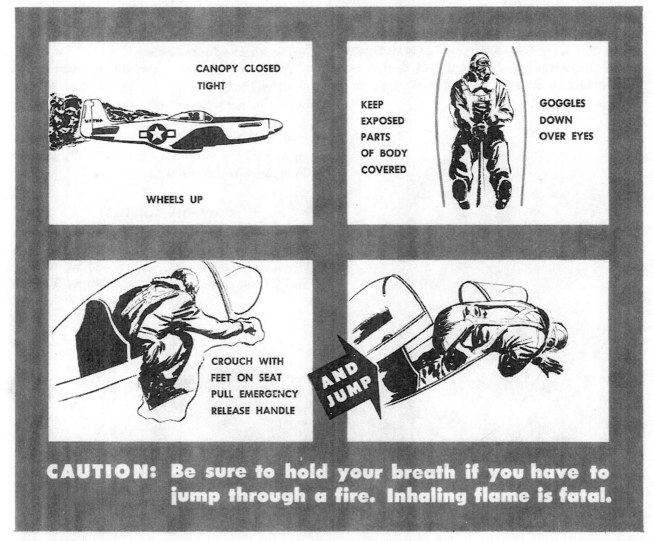

CAUTION: Be sure to hold your breath if you have to jump through a fire. Inhaling flame is fatal.

DITCHING

Never attempt to ditch the P-51 except as a last resort. Fighter planes are not designed to float on water, and the P-51 has an even greater tendency to dive because of the airscoop underneath. It will go down in 1½ to 2 seconds.

It is possible to ditch the P-51 successfully and it has been done on several occasions. However, it is a hazardous business.

If trouble arises during an over-water flight, and if you're sure that you can't reach land, don't hesitate to bail out. You won't be able to save the airplane by a water landing, nor will you provide yourself with any useful equipment, as would be the case with a larger airplane. So you had best abandon ship in the air.

If you're at an extremely low altitude, pull the airplane up at a steep angle, throw off the safety belt and shoulder straps and go out over the right side. Even if you're flying only 50 feet above the water, or less, you have sufficient speed at the minimum cruising rate to pull up to 500 feet, and from that altitude you can make a safe jump.

The important thing to remember is to make as steep a pull-up as possible and get out at as high an altitude as you can.

If there is a fire, or if for any other reason it is advisable to go out on the left side rather than the right side, don't hesitate to do so. The right side is recommended only because the slipstream helps you in clearing the airplane.

If it isn't possible to get up high enough to make a successful parachute drop, remember that the P-51 can be ditched successfully.

Radio Procedure

Theoretically, the radio procedure that you're to follow before ditching is the same as that for an over-water bailout, described on page 88. Accomplish as much of it as time and circumstances will permit. Your chances of a speedy rescue will depend heavily on whether an Air/Sea Rescue Unit can get a good fix on your position.

Approach and Touchdown

You can make an approximate estimate of wind velocity from the appearance of the water, in accordance with the table below. If the wind is less than 35 mph, touch down parallel with the lines of wave crests and troughs. Ditch into the wind only if its velocity is over 35 mph, or if the sea is flat.

WIND VELOCITIES

A few white caps	10 to 20 mph
Many white caps	20 to 30 mph
Streaks of foam	30 to 40 mph
Spray from crests	40 to 50 mph

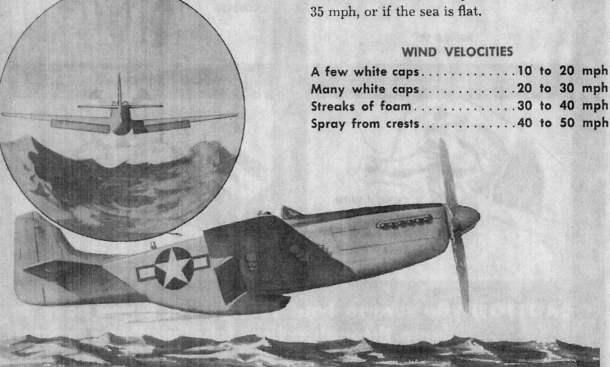

Keep the wheels up, and use flaps in proportion to available power in order to obtain minimum forward speed with minimum rate of descent. Approach in 3-point attitude, and observe the following procedure:

1. Lower the seat, duck your head, and jettison the canopy.
2. Jettison tanks or bombs, if you're carrying any.
3. Unfasten the parachute harness.
4. Make sure that your shoulder harness and safety belt are locked and tight.
5. Maintain an airspeed of 120 mph.
6. Cut the switches just before impact.
7. Touch down in normal landing attitude. Deceleration following contact will be very violent.

Once the airplane stops you won't have more than 2 seconds, so fix in your mind the following procedure:

1. Release the safety belt.
2. Jump out and pull the life raft loose from the parachute.
3. Inflate your Mae West immediately after discarding your parachute harness.
4. Inflate the life raft and wriggle into it.

Even in shallow water don't get out of your Mae West until you are safely on shore. Also remember to salvage your parachute, if possible. It may come in handy.

BAILOUT PROCEDURE

There are several successfully tried and tested methods of bailing out of the P-51, when the airplane is under control. However, the following bailout procedure is recommended, since it remains essentially the same whether the airplane is under control, on fire, or in a spin.

1. Slow the airplane to the lowest speed that is reasonably safe—usually about 150 mph. The lower the speed at which you bail out, the less risk there is. But don't slow the airplane dangerously near the stalling point, particularly if you have no power.
2. Lower the seat, duck your head, and jettison the canopy.
3. Disconnect your headset and oxygen hose, and release the safety belt and shoulder harness. Make a special point of throwing the straps off your shoulders. This is important because, even if the straps are loose, it is more difficult to rise with the straps hanging on your shoulders.
4. Pull yourself up onto the seat so that you're in a crouching position with your feet in the seat.
5. Dive with head down toward the trailing edge of the right wing, unless a fire or some other condition makes it advisable to go out the left side. The right side is recommended because the slipstream helps you in clearing the airplane. However, don't hesitate to go out the left side, for successful jumps have been made from both sides of the airplane.

JUMP TOWARDS WING FROM EITHER SIDE OF PLANE

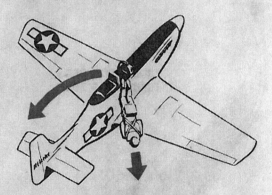

If you get in trouble at high altitude, bring the plane to a lower altitude before bailing out, when possible. But if you must jump from a high altitude, your best bet is to make a delayed free fall before opening your chute. In this way you not only escape the danger of cold, of lack of oxygen, and—if in a combat zone—the danger of gunfire, but you also eliminate the possibility of personal injury from the snap-out in the rarified air. At high altitudes the G force exerted on the pilot by the pull of the harness in the opening snap is from two to four times as great as at lower altitudes.

Before bailing out at high altitudes, open the emergency knob on the oxygen regulator and fill your lungs with oxygen. Take several good breaths and then hold your breath as long as possible. If you do this before a delayed free fall, you have a supply of oxygen to help you until you reach the lower altitudes.

Bailing Out of a Spin

If you ever have to bail out of a spin, it is recommended that you jump to the **inside**, not the outside of the spin.

If the airplane is spinning to the right, go out over the right side of the cockpit, if it's spinning to the left, go out over the left side. Following this procedure, you have a much better chance of falling clear of the airplane and escaping the centrifugal force of the spin, which tends to throw you into the tail assembly or the prop as it comes around.

Records compiled by the Office of Flying Safety show that of all the successful jumps made out of spins in all types of aircraft, by far the greatest percentage was made from **inside** the spin.

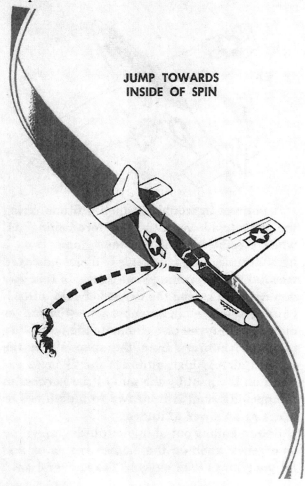

JUMP TOWARDS INSIDE OF SPIN

Bailing Out Over Water

If you have to bail out over water, it's extremely important that you follow a definite radio procedure, to give you the best possible chance of being picked up in a hurry. Each theater may have its own radio procedure—you'll get the full details when briefed for a mission.

If there is opportunity and time, try to gain altitude, particularly if you're below 5000 feet. Doing so increases the range of your VHF transmission and help Air/Sea Rescue Units get a good fix. How quickly you are rescued may be determined by the accuracy of this fix. Given below is a typical radio procedure for bailout over water or for ditching.

1. Notify wingmen that your airplane is in trouble.
2. If your airplane is equipped with IFF, turn the emergency switch ON.
3. Transmit "Mayday" three times, followed by the call sign of your plane three times.
4. Your first transmission will be on the assigned air-ground frequency. If you cannot establish communication on this frequency, use any other available frequency in an effort to establish contact with a ground station.
5. If time permits, give the following information:
   a. Estimated position and time
   b. Course and speed
   c. Altitude
   d. Your intention as to bailing out or ditching.
   e. Just before bailing out, break the safety wire on the VHF control switch and throw the switch to TR.
   f. In case your trouble becomes remedied and you do not have to bail out, a message cancelling the distress signal must be sent out on the same frequency.

Wingmen or flight members, on hearing your call on operations channel, should if possible orbit the spot, one plane going down low, another remaining high and continuing transmission of distress signals. This insures that a good fix is obtained.

For further information on bailout procedures, water landings, and control of parachutes, see your PIF.

INSTRUMENT FLYING

Check your vacuum gage readings for 3.75—4.25 inches and cross-check all of your instruments frequently

Instrument training in a single-place fighter plane is naturally limited. You are confined pretty much to such artificial aids as the Link and slow-speed 2-place planes. Since these facilities are of only limited value, some instrument training in fighter planes is done with the pilot flying blind and an observation plane alongside to prevent any difficulties should the pilot on instruments cross an airway or get into a swarm of PT's or other training planes.

Although instrument training in single-place fighters is rather difficult, the importance of instrument flying in these airplanes cannot be over-emphasized. If instrument flying gets you out of a tight situation just once, it is worth all the time spent in learning it. Because radio navigation aids cannot be widely used in combat zones, in some theaters a knowledge of instrument flying is absolutely necessary.

RESTRICTED

It's not within the scope of this manual to give instructions on instrument flight—that's taken care of in your training. But the information below is worth remembering, since it is derived from data of flight tests made to determine the P-51D's instrument flight characteristics.

Altitude Control

Your rate of climb or descent, at a given airspeed and power setting, is determined by the degree of pitch, or nose attitude change. This wouldn't be of much importance if you were flying a slow airplane. However, when you're cruising at 250 to 300 mph, a very slight change of nose attitude will immediately result in a high rate of climb or descent, with a rapid gain or loss of altitude. Therefore, when you are maneuvering at low altitude under instrument conditions, as during an instrument approach, the primary rule of safety is: **Keep your airspeed down.**

As a matter of fact, your precision in instrument flight maneuvers at any altitude will be greatly increased if you cut back to slow cruising speed.

Bank Control

The turn needle is gyro-actuated and indicates rate of turn only, regardless of speed. Therefore, at a given rate of turn, the angle of bank in a coordinated turn depends upon true airspeed. A standard-rate turn at an altitude of 1000 feet and an IAS of 200 mph will require approximately 27° of bank. But at 25,000 feet an IAS of 200 mph will require about 37° of bank to accomplish a standard-rate turn, because the TAS at that altitude is in excess of 300 mph.

Control pressure on the elevators changes rapidly during the entry into a steeply banked turn, and it's very easy at this time to make inadvertent changes in your pitch, or nose attitude. As explained above, these slight changes in nose attitude, at high speeds, will result in large altitude variations; these can be critically dangerous if you're on instruments and close to the ground.

This hazard, too, can be largely avoided by slowing the airplane. When you cut speed, the angle of bank required for a given rate of turn is greatly lessened and the problem of control is proportionately reduced.

Control Sensitivity

Owing to the sensitivity of your P-51's controls, it is essential that you remain mentally attentive to the instruments at all times. Accurate trim control is extremely important; it will contribute greatly to your physical relaxation; and allow you to concentrate on the numerous unrelated details of instrument flight.

TRUE AIRSPEED	100 MPH	150 MPH	200 MPH	300 MPH	400 MPH
ANGLE OF BANK FOR 3°/SEC. TURN	13.5°	19.8°	25.6°	35.75°	43.8°
ANGLE OF BANK FOR 1½°/SEC. TURN	6.85°	10.2°	13.5°	19.8°	25.6°

RESTRICTED

So, when you're on instruments, trim carefully and as often as required.

Attitude Instrument Flying

In instrument flying today, the attitude system is in general use. Attitude instrument flying is the controlling of an airplane by reference to its attitude in relation to the horizon as evidenced by instrument indications. As in contact flight, proper attitude is attained by the coordinated use of all controls.

If a pilot is thoroughly familiar with this system of instrument flying, he should have no difficulty in continuing instrument flight even though some of his instruments are lost because of mechanical failure or enemy action. This system is explained thoroughly and at length in T. O. 30-100A-2, dated 15 December 1944. Study it and practice what it preaches until you have acquired a real proficiency in attitude instrument flying. It may save your life.

Instrument Approach

Shortly before reaching the station on the initial approach, reduce your speed to 150 mph indicated, and lower flaps 10°. This low airspeed simplifies radio procedures and increases your control of the airplane.

After completion of the initial approach, execute your final approach at 130 mph indicated, with landing gear down and flaps lowered 15°.

Although final approach speed depends largely on ceiling condition, 130 mph with 15° of flaps is recommended. It's high enough so that, if you have to go around, an emergency pull-up will be quite safe. And when you have sighted the field, this speed is low enough to permit you to lower flaps and round out for touchdown without danger of floating too far down the runway.

With the airspeeds and flap settings recommended above, standard rate turns are safe and should be used whenever practicable.

Always use flaps during an instrument approach. Doing so reduces stalling speed, and the slightly increased drag necessitates more power, which in turn improves your rudder and elevator control. Also, the lowered nose attitude improves visibility when you break contact on final approach.

TIPS ON INSTRUMENT FLYING

1. If your P-51 is equipped with a K-14 gunsight, your view of the instrument panel will probably be partially obstructed. In this case, lower the seat. You'll still have enough forward vision to land the airplane, so it won't be necessary to raise the seat again during the last few seconds of final approach.

2. When you encounter severe turbulence, adjust the power settings to give you approximately 160-180 mph IAS. Don't try to maintain one definite airspeed. Set the throttle and prop control, and then let the airspeed oscillate in a 20 mph range. Doing so will give you complete control of the airplane without risk of structural damage.

3. Always remember that as the airspeed increases, the degree of bank increases for a given rate of turn. At high airspeeds a standard rate of turn may require a dangerously steep angle of bank. In view of this fact, your rate of turn should never be such as to require an angle of bank greater than you can safely control.

NIGHT FLYING

The lighting equipment for flying the P-51 at night is as follows:

Outside the Plane

1. Wing and tail position lights. Controlled by 3-position switches (BRIGHT-OFF-DIM) on the right switch panel; one for the tail light, one for the wing lights.

2. Landing light. Swings out from the left wheel well when the landing gear is extended. Controlled by an ON-OFF switch on the coolant switch panel.

3. Identification lights. Red, amber, and green lights under the right wing-tip. The lights can be used in any combination, either steadily or intermittently. Controlled by individual 3-position switches (STEADY-OFF-KEY) on the right switch panel. The key for intermittent (code) operation is on top of a small box above the right switch panel.

Inside the Cockpit

1. Cockpit lights. Two small white lights for map reading, checking fuel gages, or whenever light is needed in the cockpit. Controlled by an ON-OFF twist switch built into the housing of each light. The cockpit light switch on the front switch panel must be ON before the lights can be used.

2. Fluorescent lights. Two barrel-shaped fluorescent lights, one on each side of the cockpit, for lighting the luminous painted dials. Controlled by individual rheostat switches: the left switch on the coolant switch panel, the right one on the right switch panel. You turn the rheostat knobs to START until the lights go on, then turn them back to the desired brightness.

To make the dials glow without any glare, the opening on the front of the housing of the lights should be **closed completely**. The dials are then illuminated by invisible ultra-violet light. Open the lights only when required for checking the position of switches and levers.

When opened, a bluish light illuminates the whole cockpit, and this causes a glare that is tiring to the eyes.

3. Flashlight. Planes are not usually equipped with a flashlight, but it is wise to carry one in case any cockpit lights burn out. If the light is equipped with a red lens, or is covered with red cellophane, you can use it freely without danger of being dazzled.

Other Lights

Also inside the cockpit are other lights which operate automatically, day or night (in most cases it is possible to dim these lights for night flying):

1. Lights on VHF control box. Indicator lights, one for each of the four channels and another to show when you are transmitting. A small slide covering can be moved over these lights to dim out everything but small indicator letters (T, and A, B, C, or D).

2. Landing gear indicator lights located on the lower part of the instrument panel. Green light glows when gear is down and locked, red when it is not. These lights may be turned to dim or brighten them.

3. Supercharger light. An amber light next to the supercharger switch to show when the supercharger is in high blower.

RESTRICTED

FLIGHT OPERATION CHARTS

An important problem in planning any long flight, whether a cross-country trip or a combat mission, is calculating the amount of fuel required and planning the best use of that fuel. This is of course especially important on long over-water hops where the difference between getting back to your base or ending up in a dinghy may depend upon nothing more than how well you plan your rpm and throttle settings.

On most combat missions, fuel and flight planning problems are worked out for you by the squadron operations officer. But every pilot should be able to work these problems out for himself and know how the charts are used.

The charts used in flight planning include:
(1) Takeoff, Climb, and Landing charts, and
(2) Flight Operation charts (for constant and level flight).

Sample charts—one of each type—are given on this and the next page. These are provided so that you can follow the procedure of solving the typical problem given below. A complete set of full-size charts will be found on the pages immediately following.

By means of these charts it is possible to approximate all the details of a complete trip.

Here is an Example

You are going on a 1000-mile hop in a P-51D equipped with a V-1650-7 engine. The plane has wing racks but you will not be carrying any external fuel load. With wing and fuselage tanks full, you will be carrying 269 gallons.

According to weather reports the most favorable winds are at 10,000 feet. You are to take off and climb to 10,000 feet above the field, then fly the 1000-mile hop.

The first step in planning the flight is to select the charts suitable for the airplane's load condition.

Takeoffs and Landings

The takeoff and landing charts can be used in determining runway distances required.

For example, in working out the cross-country problem given above, you might want to know how much concrete runway is required to make a landing at a sea level airport. Note opposite the 8000 weight figure under "Hard

AIRCRAFT MODEL (S) P-51D AND P-51K		TAKE-OFF, CLIMB & LANDING CHART																	ENGINE MODEL (S) V-1650-7	

TAKE-OFF DISTANCE FEET

GROSS WEIGHT LB.	HEAD WIND		HARD SURFACE RUNWAY						SOD-TURF RUNWAY						SOFT SURFACE RUNWAY					
			AT SEA LEVEL		AT 3000 FEET		AT 6000 FEET		AT SEA LEVEL		AT 3000 FEET		AT 6000 FEET		AT SEA LEVEL		AT 3000 FEET		AT 6000 FEET	
	M.P.H.	KTS.	GROUND RUN	TO CLEAR 50' OBJ.	GROUND RUN	TO CLEAR 50' OBJ.	GROUND RUN	TO CLEAR 50' OBJ.	GROUND RUN	TO CLEAR 50' OBJ.	GROUND RUN	TO CLEAR 50' OBJ.	GROUND RUN	TO CLEAR 50' OBJ.	GROUND RUN	TO CLEAR 50' OBJ.	GROUND RUN	TO CLEAR 50' OBJ.	GROUND RUN	TO CLEAR 50' OBJ.
11,000	0	0	1800	2700	2000	3000	2300	3300	2000	2800	2100	3100	2400	3400	2300	3200	2500	3400	2800	3900
	17	15	1400	2100	1500	2300	1800	2600	1500	2200	1600	2400	1900	2700	1700	2500	1900	2700	2200	3100
	34	30	1000	1600	1100	1800	1300	2100	1100	1700	1200	1900	1400	2100	1200	1900	1400	2100	1600	2400
	51	45	700	1200	800	1300	900	1500	700	1200	800	1400	1000	1600	800	1300	1000	1500	1200	1700
10,000	0	0	1600	2400	1800	2500	2000	2800	1700	2400	1800	2600	2100	3000	1900	2700	2100	2900	2400	3200
	17	15	1200	1800	1300	2000	1500	2300	1300	1900	1400	2100	1600	2400	1400	2100	1600	2300	1800	2500
	34	30	900	1400	1000	1500	1100	1800	900	1400	1000	1600	1200	1800	1000	1600	1200	1700	1300	2000
	51	45	600	1000	700	1100	800	1300	600	1000	700	1200	800	1300	700	1100	800	1200	900	1500
9000	0	0	1400	2000	1500	2200	1700	2500	1400	2100	1600	2300	1800	2500	1600	2300	1800	2500	2000	2800
	17	15	1000	1600	1200	1700	1300	2000	1100	1600	1200	1800	1400	2000	1200	1800	1300	1900	1500	2200
	34	30	700	1200	800	1300	1000	1500	800	1200	900	1400	1000	1500	900	1300	1000	1400	1100	1700
	51	45	500	800	600	1000	700	1100	500	900	600	1000	700	1100	600	900	700	1000	800	1200

NOTE: INCREASE CHART DISTANCES AS FOLLOWS: 75°F + 10%; 100°F + 20%; 125°F + 30%; 150°F + 40%
DATA AS OF 8-20-44   BASED ON: FLIGHT TESTS
OPTIMUM TAKE-OFF WITH 3000 RPM, 61 IN. HG. & 20 DEG. FLAP IS 80% OF CHART VALUES

CLIMB DATA

GROSS WEIGHT LB.	AT SEA LEVEL				AT 5000 FEET					AT 10,000 FEET					AT 15,000 FEET					AT 20,000 FEET					AT 25,000 FEET				
	BEST I.A.S.		RATE OF CLIMB F.P.M.	GAL. OF FUEL USED	BEST I.A.S.		RATE OF CLIMB F.P.M.	FROM SEA LEVEL		BEST I.A.S.		RATE OF CLIMB F.P.M.	FROM SEA LEVEL		BEST I.A.S.		RATE OF CLIMB F.P.M.	FROM SEA LEVEL		BEST I.A.S.		RATE OF CLIMB F.P.M.	FROM SEA LEVEL		BEST I.A.S.		RATE OF CLIMB F.P.M.	FROM SEA LEVEL	
	MPH	KTS			MPH	KTS		TIME MIN.	FUEL USED	MPH	KTS		TIME MIN.	FUEL USED	MPH	KTS		TIME MIN.	FUEL USED	MPH	KTS		TIME MIN.	FUEL USED	MPH	KTS		TIME MIN.	FUEL USED
11,000	175	150	1450	15	175	150	1500	3.4	20	175	150	1500	6.8	26	170	150	1450	10.5	31	165	145	1150	14.0	37	165	145	1100	18.5	44
10,000	175	150	1750	15	175	150	1750	2.8	20	175	150	1800	5.6	24	170	150	1800	8.5	29	165	145	1450	11.5	34	165	145	1400	15.0	40
9000	175	150	2050	15	175	150	2100	2.4	19	175	150	2150	4.8	23	170	150	2150	7.5	27	165	145	1850	9.5	31	165	145	1800	12.5	36

POWER PLANT SETTINGS (DETAILS ON FIG. 34, SECTION III):
DATA AS OF 8-20-44   BASED ON: FLIGHT TESTS
FUEL USED (U.S. GAL.) INCLUDES WARM-UP & TAKE-OFF ALLOWANCE

LANDING DISTANCE FEET

GROSS WEIGHT LB.	BEST IAS APPROACH				HARD DRY SURFACE						FIRM DRY SOD						WET OR SLIPPERY					
	POWER OFF		POWER ON		AT SEA LEVEL		AT 3000 FEET		AT 6000 FEET		AT SEA LEVEL		AT 3000 FEET		AT 6000 FEET		AT SEA LEVEL		AT 3000 FEET		AT 6000 FEET	
	MPH	KTS	MPH	KTS	GROUND ROLL	TO CLEAR 50' OBJ.	GROUND ROLL	TO CLEAR 50' OBJ.	GROUND ROLL	TO CLEAR 50' OBJ.	GROUND ROLL	TO CLEAR 50' OBJ.	GROUND ROLL	TO CLEAR 50' OBJ.	GROUND ROLL	TO CLEAR 50' OBJ.	GROUND ROLL	TO CLEAR 50' OBJ.	GROUND ROLL	TO CLEAR 50' OBJ.	GROUND ROLL	TO CLEAR 50' OBJ.
10,000	130	115	130	115	1300	2500	1600	2600	1700	2800	1500	2600	1700	2800	1900	3000	3500	4800	3900	5100	4400	5500
9000	130	115	130	115	1200	2300	1400	2400	1500	2600	1400	2400	1600	2600	1700	2800	3200	4300	3500	4600	3900	5000
8000	130	115	130	115	1100	2100	1200	2200	1400	2400	1300	2200	1400	2400	1500	2500	2900	4000	3100	4100	3400	4500

DATA AS OF 8-20-44   BASED ON: FLIGHT TESTS
OPTIMUM LANDING IS 80% OF CHART VALUES

REMARKS:
NOTE: TO DETERMINE FUEL CONSUMPTION IN BRITISH IMPERIAL GALLONS, MULTIPLY BY 10, THEN DIVIDE BY 12.

LEGEND
I.A.S. = INDICATED AIRSPEED
M.P.H. = MILES PER HOUR
KTS. = KNOTS
F.P.M. = FEET PER MINUTE

dry surface, at sea level" that 2100 feet is required to clear a 50 foot object.

Planning the Flight

The Flight Operation chart selected shows that a P-51D with no external tanks weighs 8000-9600 pounds (see top center of chart). Therefore on the Climb chart opposite the 9000 weight figure and under the 10,000 feet altitude figure, you find that the fuel used in the warm-up, takeoff and climb to 10,000 feet will be 23 gallons. (Notice also the underlined figures which show that the best climbing speed is 175 IAS, the rate of climb is 2150 feet per minute, and the time for the climb is 4.8 minutes.)

Deducting 23 gallons from the original supply of 269 leaves 246. Therefore on the Flight Operation chart opposite 240 (the nearest figure) find the range nearest (but not below) that of your trip. In this case it is 1100 miles.

Dropping below this 1100 figure and opposite the 10,000 feet altitude figure you find this information for the flight:

Rpm setting . . . . 2200 rpm
Manifold pressure . . . . 40 inches
Mixture setting . . . . RUN
Supercharger . . . . Low blower
Fuel consumption . . . . 71 gal/hour
Speed . . . . 325 TAS

To travel 1000 miles at 325 TAS will require 3.1 hours (1000 divided by 325). At 71 gallons per hour you will therefore use 220 gallons (71 times 3.1) in flying to your destination. Since you had 246 gallons remaining after the takeoff and climb, you will have 26 gallons (246 minus 220) left for the letdown and landing—enough for about 30 minutes of additional flying time.

Remember:

Figures are approximate. Always allow a margin of safety.

FLIGHT OPERATION INSTRUCTION CHART

AIRCRAFT MODEL(S): P-51D AND P-51K
ENGINE(S): V-1650-7

CHART WEIGHT LIMITS: 10,600 TO 10,000 POUNDS

EXTERNAL LOAD ITEMS
1 - 500-POUND OR SMALLER WING BOMBS
NUMBER OF ENGINES OPERATING: 1

LIMITS	RPM	M.P. IN.HG.	BLOWER POSITION	MIXTURE	TIME LIMIT MIN.	CYL. TEMP.	TOTAL G.P.H.
WAR EMERG.	3000	67	LOW	RUN	5		211
			HIGH	RUN	5		215.
MILITARY POWER	3000	61	LOW	RUN	15		182
			HIGH	RUN	15		187

INSTRUCTIONS FOR USING CHART: SELECT FIGURE IN FUEL COLUMN EQUAL TO OR LESS THAN AMOUNT OF FUEL TO BE USED FOR CRUISING(1). MOVE HORIZONTALLY TO RIGHT OR LEFT AND SELECT RANGE VALUE EQUAL TO OR GREATER THAN THE STATUTE OR NAUTICAL AIR MILES TO BE FLOWN. VERTICALLY BELOW AND OPPOSITE VALUE NEAREST DESIRED CRUISING ALTITUDE (ALT.) READ RPM, MANIFOLD PRESSURE (M.P.) AND MIXTURE SETTING REQUIRED.

NOTES: COLUMN I IS FOR EMERGENCY HIGH SPEED CRUISING ONLY. COLUMNS II, III, IV AND V GIVE PROGRESSIVE INCREASE IN RANGE AT A SACRIFICE IN SPEED. AIR MILES PER GALLON (MI./GAL.) (NO WIND), GALLONS PER HR. (G.P.H.) AND TRUE AIRSPEED (T.A.S.) ARE APPROXIMATE VALUES FOR REFERENCE. RANGE VALUES ARE FOR AN AVERAGE AIRPLANE FLYING ALONE (NO WIND). TO OBTAIN BRITISH IMPERIAL GAL. (OR G.P.H.) MULTIPLY U.S. GAL. (OR G.P.H.) BY 10 THEN DIVIDE BY 12.

COLUMN I			COLUMN II			COLUMN III			COLUMN IV			COLUMN V		
RANGE IN AIRMILES		FUEL U.S. GAL.	RANGE IN AIRMILES		(3.15 NAUT.) MI./GAL.	RANGE IN AIRMILES		(3.5 NAUT.) MI./GAL.	RANGE IN AIRMILES		(3.9 NAUT.) MI./GAL.	RANGE IN AIRMILES		FUEL U.S. GAL.
STATUTE	NAUTICAL		STATUTE	NAUTICAL		STATUTE	NAUTICAL		STATUTE	NAUTICAL		STATUTE	NAUTICAL	
910	790	269	970	840		1090	950		1220	1060		1340	1160	269
810	700	240	860	750		980	850		1090	950		1190	1040	240
740	640	220	790	690		900	780		1000	870		1100	950	220
670	580	200	720	630		820	710		910	790		1000	860	200
600	530	180	640	560		740	640		820	720		900	780	180
540	470	160	570	500		650	570		730	640		800	690	160

MAXIMUM CONTINUOUS / **MAXIMUM AIR RANGE**

R.P.M.	M.P. INCHES	MIX. TURE	TOT. GPH	APPROX. T.A.S. MPH / KTS	PRESS ALT. FEET	R.P.M.	M.P. INCHES	MIX. TURE	TOT. GPH	APPROX. T.A.S. MPH / KTS	R.P.M.	M.P. INCHES	MIX. TURE	TOT. GPH	APPROX. T.A.S. MPH / KTS	R.P.M.	M.P. INCHES	MIX. TURE	TOT. GPH	APPROX. T.A.S. MPH / KTS	R.P.M.	M.P. INCHES	MIX. TURE	TOT. GPH	APPROX. T.A.S. MPH / KTS
			98	375 325	40000																				
			94	355 310	35000																				
			103	355 310	30000																				
					25000											2300	F.T.	RUN	75	340 295	2100	F.T.	RUN	60	295 255
					20000											2150	38	RUN	71	320 275	2150	F.T.	RUN	58	285 245
					15000	2600	44	RUN	96	350 305	2500	F.T.	RUN	89	365 315	2150	F.T.	RUN	66	300 260	1850	F.T.	RUN	54	265 230
2700	46	RUN	98	330 290	10000	2600	44	RUN	90	325 280	2450	RUN	RUN	84	340 295	2000	38	RUN	62	280 240	1650	34	RUN	49	240 210
2700	46	RUN	92	310 270	5000	2600	44	RUN	84	300 260	2350	RUN	RUN	79	325 280	1950	38	RUN	57	260 225	1600	34	RUN	45	220 190
2700	46	RUN	86	290 255	S.L.	2600	44	RUN	78	280 245	2250	RUN	RUN	75	300 260	1950	38	RUN	57	260 225	1600	33	RUN	42	200 175
											2250	RUN	RUN	69	280 245	1950	38	RUN	53	240 205					
											2250	RUN	RUN	64	260 225										

SPECIAL NOTES

(1) MAKE ALLOWANCE FOR WARM-UP, TAKE-OFF & CLIMB (SEE FIG.53.) PLUS ALLOWANCE FOR WIND, RESERVE AND COMBAT AS REQUIRED.

HIGH BLOWER ABOVE HEAVY LINE

FOR DETAILS SEE POWER PLANT CHART (FIG. 34, SECT. III)

EXAMPLE

AT 10,400 LB. GROSS WEIGHT WITH 230 GAL. OF FUEL (AFTER DEDUCTING TOTAL ALLOWANCES OF 39 GAL.) TO FLY 1000 STAT. AIRMILES AT 15,000 FT. ALTITUDE MAINTAIN 2150 RPM AND F.T. IN. MANIFOLD PRESSURE WITH MIXTURE SET: RUN

LEGEND

ALT. : PRESSURE ALTITUDE
M.P. : MANIFOLD PRESSURE
GPH : U.S. GAL. PER HOUR
TAS : TRUE AIRSPEED
KTS. : KNOTS
S.L. : SEA LEVEL

F.R. : FULL RICH
A.R. : AUTO-RICH
A.L. : AUTO-LEAN
C.L. : CRUISING LEAN
M.L. : MANUAL LEAN
F.T. : FULL THROTTLE

DATA AS OF: 9-10-44 BASED ON: FLIGHT TESTS

FLIGHT OPERATION INSTRUCTION CHART

AIRCRAFT MODEL(S): P-51D AND P-51K
ENGINE(S): V-1650-7

CHART WEIGHT LIMITS: 10,000 TO 8,500 POUNDS

EXTERNAL LOAD ITEMS
2 - 500-POUND OR SMALLER WING BOMBS
NUMBER OF ENGINES OPERATING: 1

LIMITS	M.P. IN. HG.	BLOWER POSITION	MIXTURE	TIME LIMIT	CYL. TEMP.	TOTAL G.P.H.
WAR EMERG.	67	LOW	RUN	5 MIN.		211
		HIGH	RUN	5 MIN.		215
MILITARY POWER	61	LOW	RUN	15 MIN.		182
		HIGH	RUN	15 MIN.		187

INSTRUCTIONS FOR USING CHART: SELECT FIGURE IN FUEL COLUMN EQUAL TO OR LESS THAN AMOUNT OF FUEL TO BE USED FOR CRUISING. MOVE HORIZONTALLY TO RIGHT OR LEFT AND SELECT RANGE VALUE EQUAL TO OR GREATER THAN THE STATUTE OR NAUTICAL AIR MILES TO BE FLOWN. VERTICALLY BELOW AND OPPOSITE VALUE NEAREST DESIRED CRUISING ALTITUDE (ALT.) READ RPM, MANIFOLD PRESSURE (M.P.) AND MIXTURE SETTING REQUIRED.

NOTES: COLUMN I IS FOR EMERGENCY HIGH SPEED CRUISING ONLY. COLUMNS II, III, IV AND V GIVE PROGRESSIVE INCREASE IN RANGE AT A SACRIFICE IN SPEED. AIR MILES PER GALLON (MI./GAL.) (NO WIND), GALLONS PER HR. (G.P.H.) AND TRUE AIRSPEED (T.A.S.) ARE APPROXIMATE VALUES FOR REFERENCE. RANGE VALUES ARE FOR AN AVERAGE AIRPLANE FLYING ALONE (NO WIND). TO OBTAIN BRITISH IMPERIAL GAL (OR G.P.H.) MULTIPLY U.S. GAL. (OR G.P.H.) BY 10 THEN DIVIDE BY 12.

COLUMN I — (3.6 STAT. (3.15 NAUT.) MI./GAL.)

RANGE IN AIRMILES		FUEL	RANGE IN AIRMILES	
STATUTE	NAUTICAL	U.S. GAL.	STATUTE	NAUTICAL
620	540	184	660	580
540	470	160	570	500
470	410	140	500	440
400	350	120	430	380
340	290	100	360	310
270	230	80	280	250
200	170	60	210	190
130	110	40	140	120
60	60	20	70	60

PRESS ALT. FEET	R.P.M.	M.P. INCHES	MIX-TURE	TOT. GPH	T.A.S. MPH	T.A.S. KTS
40000						
35000						
30000						
25000	2700	46	RUN	98	375	325
20000	2700	46	RUN	94	355	310
15000	2700	46	RUN	103	355	310
10000	2700	46	RUN	98	330	290
5000	2700	46	RUN	92	310	270
S.L.	2700	46	RUN	86	290	255

COLUMN II

RANGE IN AIRMILES		FUEL	
STATUTE	NAUTICAL		
580	500		SUBTRACT FUEL ALLOWANCES NOT AVAILABLE FOR CRUISING
440	380		
310	250		
190	120		
	60		

R.P.M.	M.P. INCHES	MIX-TURE	TOT. GPH	T.A.S. MPH	T.A.S. KTS
2600	44	RUN	96	350	305
2600	44	RUN	90	325	280
2550	44	RUN	84	300	260
		RUN	78	280	245

COLUMN III — (4.1 STAT. (3.55 NAUT.) MI./GAL.)

RANGE IN AIRMILES	
STATUTE	NAUTICAL
750	650
650	570
570	490
490	420
410	350
320	280
240	210
180	140
80	70

R.P.M.	M.P. INCHES	MIX-TURE	TOT. GPH	T.A.S. MPH	T.A.S. KTS
2500	43	RUN	89	365	315
2450	43	RUN	83	310	295
2350	F.T.	RUN	79	325	280
2250	41	RUN	75	300	260
2250	41	RUN	69	280	245
2250	40	RUN	63	260	225

COLUMN IV — (4.6 STAT. (4 NAUT.) MI./GAL.)

RANGE IN AIRMILES	
STATUTE	NAUTICAL
840	730
730	640
640	560
550	480
460	400
370	320
270	240
180	160
90	80

R.P.M.	M.P. INCHES	MIX-TURE	TOT. GPH	T.A.S. MPH	T.A.S. KTS
2250	38	F.T.	73	335	290
2150	38	F.T.	69	315	275
2150	F.T.	RUN	65	300	260
2000	37	RUN	59	275	240
1900	37	RUN	55	255	220
1850	37	RUN	51	235	205

COLUMN V — MAXIMUM AIR RANGE

RANGE IN AIRMILES	
STATUTE	NAUTICAL
920	800
800	700
700	610
600	520
500	430
400	350
300	260
200	170
100	80

R.P.M.	M.P. INCHES	MIX-TURE	TOT. GPH	T.A.S. MPH	T.A.S. KTS
2050	F.T.	RUN	58	290	250
2100	F.T.	RUN	56	280	245
1850	F.T.	RUN	53	265	230
1650	32	RUN	48	240	210
1600	32	RUN	44	220	190
1600	31	RUN	40	200	175

MAXIMUM CONTINUOUS

SPECIAL NOTES

(1) MAKE ALLOWANCE FOR WARM-UP, TAKE OFF & CLIMB (SEE FIG. 53.) PLUS ALLOWANCE FOR WIND, RESERVE AND COMBAT AS REQUIRED.

HIGH BLOWER ABOVE HEAVY LINE

EXAMPLE

AT 9600 LB. GROSS WEIGHT WITH 150 GAL. OF FUEL (AFTER DEDUCTING TOTAL ALLOWANCES OF 34 GAL.) TO FLY 500 STAT. AIRMILES AT 15,000 FT. ALTITUDE MAINTAIN 2150 RPM AND F.T. IN. MANIFOLD PRESSURE WITH MIXTURE SET: RUN

LEGEND
- ALT.: PRESSURE ALTITUDE
- M.P.: MANIFOLD PRESSURE
- GPH: U.S. GAL. PER HOUR
- TAS: TRUE AIRSPEED
- KTS: KNOTS
- S.L.: SEA LEVEL
- F.R.: FULL RICH
- A.R.: AUTO-RICH
- A.L.: AUTO-LEAN
- C.L.: CRUISING LEAN
- M.L.: MANUAL LEAN
- F.T.: FULL THROTTLE

DATA AS OF 9-10-44 BASED ON: FLIGHT TESTS

FLIGHT OPERATION INSTRUCTION CHART

AIRCRAFT MODEL(S): P-51D AND P-51K

ENGINE(S): V-1650-7

CHART WEIGHT LIMITS: 11,500 TO 10,600 POUNDS

EXTERNAL LOAD ITEMS: 2 – 1000-POUND WING BOMBS

NUMBER OF ENGINES OPERATING: 1

LIMITS	RPM	M.P. IN.HG.	BLOWER POSITION	MIXTURE POSITION	TIME LIMIT MIN.	CYL. TEMP.	TOTAL G.P.H.
WAR EMERG.	3000	67	LOW	RUN	5		211
			HIGH	RUN	5		215
MILITARY POWER	3000	61	LOW	RUN	15		182
			HIGH	RUN	15		187

INSTRUCTIONS FOR USING CHART: SELECT FIGURE IN FUEL COLUMN EQUAL TO OR LESS THAN AMOUNT OF FUEL TO BE USED FOR CRUISING. MOVE HORIZONTALLY TO RIGHT OR LEFT AND SELECT RANGE VALUE EQUAL TO OR GREATER THAN THE STATUTE OR NAUTICAL AIR MILES TO BE FLOWN. VERTICALLY BELOW AND OPPOSITE VALUE NEAREST DESIRED CRUISING ALTITUDE (ALT.) READ RPM, MANIFOLD PRESSURE (M.P.) AND MIXTURE SETTING REQUIRED.

NOTES: COLUMN I IS FOR EMERGENCY HIGH SPEED CRUISING ONLY. COLUMNS II, III, IV AND V GIVE PROGRESSIVE INCREASE IN RANGE AT A SACRIFICE IN SPEED. AIR MILES PER GALLON (MI./GAL.) (NO WIND), GALLONS PER HR. (G.P.H.) AND TRUE AIRSPEED (T.A.S.) ARE APPROXIMATE VALUES FOR REFERENCE. RANGE VALUES ARE FOR AN AVERAGE AIRPLANE FLYING ALONE (NO WIND). TO OBTAIN BRITISH IMPERIAL GAL. (OR G.P.M.) MULTIPLY U.S. GAL. (OR G.P.M.) BY 10 THEN DIVIDE BY 12.

COLUMN I (STAT. (NAUT.) MI./GAL.)

RANGE IN AIRMILES		FUEL U.S. GAL.	R.P.M.	M.P. INCHES	MIX. TURE	TOT. GPH	T.A.S. MPH	T.A.S. KTS
STATUTE	NAUTICAL							
890	770	269	3000	46	RUN	98	365	320
			3000	46	RUN	94	345	300
790	690	240	2700	46	RUN	103	345	300
730 / 660	630 / 570	220 / 200	2700	46	RUN	98	325	285
			2700	46	RUN	92	305	265
590 / 530	510 / 460	180 / 160	2700	46	RUN	86	285	250

COLUMN II (STAT. (NAUT.) MI./GAL.)

SUBTRACT FUEL ALLOWANCES NOT AVAILABLE FOR CRUISING

COLUMN III (3.7 STAT. (3.25 NAUT.) MI./GAL.)

RANGE IN AIRMILES		R.P.M.	M.P. INCHES	MIX. TURE	TOT. GPH	T.A.S. MPH	T.A.S. KTS
STATUTE	NAUTICAL						
1040	900						
930	810						
850 / 780	740 / 680						
710 / 630	610 / 550	2500	44	RUN	91	335	290
		2500	43	RUN	83	310	270
		2500	43	RUN	78	290	250
		2450	43	RUN	73	270	235

COLUMN IV (4.2 STAT. (3.65 NAUT.) MI./GAL.)

RANGE IN AIRMILES		R.P.M.	M.P. INCHES	MIX. TURE	TOT. GPH	T.A.S. MPH	T.A.S. KTS
STATUTE	NAUTICAL						
1150	1000	2400	F.T.	RUN	79	310	295
		2300	41	RUN	77	320	280
1030	900	2300	F.T.	RUN	74	310	270
950 / 860	820 / 750	2200	40	RUN	70	290	250
		2150	40	RUN	65	270	235
780 / 700	680 / 610	2150	39	RUN	60	250	215

COLUMN V

FUEL U.S. GAL.	RANGE IN AIRMILES		PRESS ALT. FEET
	STATUTE	NAUTICAL	
269	1310	1140	40000 / 35000 / 30000
240	1170	1020	25000
220 / 200	1070 / 970	930 / 850	20000
180 / 160	880 / 780	760 / 680	15000

MAXIMUM AIR RANGE

R.P.M.	M.P. INCHES	MIX. TURE	TOT. GPH	T.A.S. MPH	T.A.S. KTS
2200	F.T.	RUN	61	280	245
2000	F.T.	RUN	58	270	235
1750	36	RUN	54	250	215
1650	35	RUN	50	230	200
1600	34	RUN	46	210	180

MAXIMUM CONTINUOUS

PRESS ALT. FEET	R.P.M.	M.P. INCHES	MIX. TURE	TOT. GPH	T.A.S. MPH	T.A.S. KTS
40000 / 35000 / 30000	2700	46	RUN	98	365	320
	2700	46	RUN	94	345	300
25000	2700	46	RUN	103	345	300
20000	2700	46	RUN	98	325	285
15000						
10000	2700	46	RUN	92	305	265
5000						
S.L.	2700	46	RUN	86	285	250

SPECIAL NOTES

(1) MAKE ALLOWANCE FOR WARM-UP, TAKE-OFF & CLIMB (SEE FIG. 5) PLUS ALLOWANCE FOR WIND, RESERVE AND COMBAT AS REQUIRED.

HIGH BLOWER ABOVE HEAVY LINE

EXAMPLE

AT 11,400 LB. GROSS WEIGHT WITH 220 GAL. OF FUEL (AFTER DEDUCTING TOTAL ALLOWANCES OF 49 GAL.) TO FLY 900 STAT. AIRMILES AT 15,000 FT. ALTITUDE MAINTAIN 2300 RPM AND F.T. IN. MANIFOLD PRESSURE WITH MIXTURE SET: RUN

LEGEND

ALT.	=	PRESSURE ALTITUDE	F.R.	=	FULL RICH
M.P.	=	MANIFOLD PRESSURE	A.R.	=	AUTO-RICH
GPH	=	U.S. GAL. PER HOUR	A.L.	=	AUTO-LEAN
TAS	=	TRUE AIRSPEED	C.L.	=	CRUISING LEAN
KTS.	=	KNOTS	M.L.	=	MANUAL LEAN
S.L.	=	SEA LEVEL	F.T.	=	FULL THROTTLE

DATA AS OF 9-10-44   **BASED ON:** FLIGHT TESTS

FLIGHT OPERATION INSTRUCTION CHART

AIRCRAFT MODEL(S): P-51D AND P-51K
ENGINE(S): V-1650-7
CHART WEIGHT LIMITS: 10,600 TO 9000 POUNDS

EXTERNAL LOAD ITEMS: 2 — 1000-POUND WING BOMBS
NUMBER OF ENGINES OPERATING:

LIMITS	R.P.M.	M.P. IN. HG.	BLOWER POSITION	MIXTURE POSITION	TIME LIMIT	CYL. TEMP.	TOTAL G.P.H.
WAR EMERG.	3000	67	LOW	RUN	5 MIN.		211
			HIGH				215
MILITARY POWER	3000	61	LOW	RUN	15 MIN.		182
			HIGH				187

NOTES: COLUMN I IS FOR EMERGENCY HIGH SPEED CRUISING ONLY. COLUMNS II, III, IV AND V GIVE PROGRESSIVE INCREASE IN RANGE AT A SACRIFICE IN SPEED. AIR MILES PER GALLON (MI./GAL.) (NO WIND), GALLONS PER HR. (G.P.H.) AND TRUE AIRSPEED (T.A.S.) ARE APPROXIMATE VALUES FOR REFERENCE. RANGE VALUES ARE FOR AN AVERAGE AIRPLANE FLYING ALONE (NO WIND). TO OBTAIN BRITISH IMPERIAL GAL (or G.P.H.); MULTIPLY U.S. GAL (or G.P.H.) BY 10 THEN DIVIDE BY 12.

INSTRUCTIONS FOR USING CHART: SELECT FIGURE IN FUEL COLUMN EQUAL TO OR LESS THAN AMOUNT OF FUEL TO BE USED FOR CRUISING. MOVE HORIZONTALLY TO RIGHT OR LEFT AND SELECT RANGE VALUE EQUAL TO OR GREATER THAN THE STATUTE OR NAUTICAL AIR MILES TO BE FLOWN. VERTICALLY BELOW AND OPPOSITE VALUE NEAREST DESIRED CRUISING ALTITUDE (ALT.) READ RPM, MANIFOLD PRESSURE (M.P.) AND MIXTURE SETTING REQUIRED.

COLUMN I (3.15 NAUT.)

RANGE IN AIRMILES		FUEL U.S. GAL.
STATUTE	NAUTICAL	
600	520	184
520	450	160
450	400	140
390	340	120
320	280	100
260	220	80
190	170	60
130	110	40
60	50	20

MAXIMUM CONTINUOUS — (3.6 STAT. / 3.15 NAUT.) MI./GAL.

R.P.M.	M.P. INCHES	MIX-TURE	APPROX. T.A.S.		TOT. GPH
			MPH	KTS.	
2700	46	RUN	370	320	98
2700	46	RUN	345	300	94
2700	46	RUN	315	300	103
2700	46	RUN	325	280	98
2700	46	RUN	305	265	92
2700	46	RUN	285	250	86

COLUMN II (3.6 STAT. / 3.15 NAUT.)

RANGE IN AIRMILES		FUEL U.S. GAL.
STATUTE	NAUTICAL	
660	580	184
570	500	160
500	440	140
430	380	120
360	310	100
290	250	80
210	190	60
140	120	40
70	60	20

SUBTRACT FUEL ALLOWANCES NOT AVAILABLE FOR CRUISING (1)

PRESS. ALT. FEET	R.P.M.	M.P. INCHES	MIX-TURE	APPROX. T.A.S.		TOT. GPH
40000 / 35000 / 30000						
25000	2550	F.T.	RUN	335	290	91
20000	2550	44	RUN	315	275	86
15000	2550	44	RUN	295	255	82
10000 / 5000 / S.L.	2550	43	RUN	275	240	77

COLUMN III (4.0 STAT. / 3.5 NAUT.) MI./GAL.

RANGE IN AIRMILES		FUEL U.S. GAL.
STATUTE	NAUTICAL	
730	640	184
640	560	160
560	490	140
480	420	120
400	350	100
320	280	80
240	210	60
160	140	40
80	70	20

PRESS. ALT. FEET	R.P.M.	M.P. INCHES	MIX-TURE	APPROX. T.A.S.		TOT. GPH
25000	2500	43	RUN	360	310	90
20000	2500	43	RUN	335	285	86
15000	2400	F.T.	RUN	320	275	80
10000	2300	41	RUN	300	260	75
5000	2300	41	RUN	280	240	70
S.L.	2300	41	RUN	260	225	65

COLUMN IV (4.4 STAT. / 3.8 NAUT.) MI./GAL.

RANGE IN AIRMILES		FUEL U.S. GAL.
STATUTE	NAUTICAL	
810	700	184
700	600	160
610	530	140
530	450	120
440	380	100
350	300	80
260	230	60
170	150	40
90	70	20

PRESS. ALT. FEET	R.P.M.	M.P. INCHES	MIX-TURE	APPROX. T.A.S.		TOT. GPH
25000	2300	F.T.	RUN	335	290	75
20000	2200	F.T.	RUN	315	275	71
15000	2200	38	RUN	300	260	68
10000	2000	38	RUN	280	240	63
5000	2000	38	RUN	260	225	59
S.L.	2000	38	RUN	240	205	55

COLUMN V

RANGE IN AIRMILES		FUEL U.S. GAL.
STATUTE	NAUTICAL	
900	780	184
780	680	160
680	590	140
580	500	120
490	420	100
390	340	80
290	250	60
190	170	40
100	80	20

MAXIMUM AIR RANGE

PRESS. ALT. FEET	R.P.M.	M.P. INCHES	MIX-TURE	APPROX. T.A.S.		TOT. GPH
				MPH	KTS.	
25000	2150	F.T.	RUN	285	250	58
20000	1900	F.T.	RUN	265	230	54
15000	1650	34	RUN	245	210	50
10000	1600	33	RUN	225	195	46
5000						
S.L.	1600	32	RUN	205	180	43

EXAMPLE

AT 10,400 LB. GROSS WEIGHT WITH 150 GAL. OF FUEL (AFTER DEDUCTING TOTAL ALLOWANCES OF 34 GAL.) TO FLY 600 STAT. AIRMILES AT 15,000 FT. ALTITUDE MAINTAIN 2200 RPM AND F.T. IN MANIFOLD PRESSURE WITH MIXTURE SET: RUN

SPECIAL NOTES

(1) MAKE ALLOWANCE FOR WARM-UP, TAKE-OFF & CLIMB (SEE FIG. 53.) PLUS ALLOWANCE FOR WIND, RESERVE AND COMBAT AS REQUIRED.

HIGH BLOWER ABOVE HEAVY LINE

LEGEND

ALT. : PRESSURE ALTITUDE
M.P. : MANIFOLD PRESSURE
GPH : U.S. GAL. PER HOUR
TAS : TRUE AIRSPEED
KTS. : KNOTS
S.L. : SEA LEVEL

F.R. : FULL RICH
A.R. : AUTO-RICH
A.L. : AUTO-LEAN
C.L. : CRUISING LEAN
M.L. : MANUAL LEAN
F.T. : FULL THROTTLE

DATA AS OF 9-10-44   BASED ON: FLIGHT TESTS

FLIGHT OPERATION INSTRUCTION CHART

AIRCRAFT MODEL(S): P-51D AND P-51K
ENGINE(S): V-1650-7

CHART WEIGHT LIMITS: 10,800 TO 9600 POUNDS

EXTERNAL LOAD ITEMS: 2 – 75-GALLON COMBAT TANKS
NUMBER OF ENGINES OPERATING: 1

LIMITS

LIMITS	RPM	M.P. IN. HG.	BLOWER POSITION	MIXTURE POSITION	TIME LIMIT	CYL. TEMP.	TOTAL G.P.H.
WAR EMERG.	3000	67	LOW / HIGH	RUN / RUN	5 MIN.		211 / 215
MILITARY POWER	3000	61	LOW / HIGH	RUN / RUN	15 MIN.		182 / 187

INSTRUCTIONS FOR USING CHART: Select figure in fuel column equal to or less than amount of fuel to be used for cruising (1). Move horizontally to right or left and select range value equal to or greater than the statute or nautical air miles to be flown. Vertically below and opposite value nearest desired cruising altitude (ALT.) read RPM, manifold pressure (M.P.) and mixture setting required.

NOTES: Column I is for emergency high speed cruising only. Columns II, III, IV and V give progressive increase in range at a sacrifice in speed. Air miles per gallon (MI./GAL.) (no wind), gallons per hr. (G.P.H.) and true airspeed (T.A.S.) are approximate values for reference. Range values are for an average airplane flying alone (no wind). To obtain British Imperial Gal. (or G.P.H.) multiply U.S. GAL. (or G.P.H.) by 10 then divide by 12.

COLUMN I — (3.6 STAT.(3.15 NAUT.) MI./GAL.)

RANGE IN AIRMILES		FUEL U.S. GAL.	PRESS ALT. FEET
STATUTE	NAUTICAL		
1410 / 1350	1220 / 1170	419 / 400	40000 / 35000
1280 / 1210	1110 / 1050	380 / 360	25000
1140 / 1080	990 / 930	340 / 320	20000 / 15000
1010 / 940 / 870	880 / 820 / 760	300 / 280 / 260	10000 / 5000 / S.L.

MAXIMUM CONTINUOUS

R.P.M.	M.P. INCHES	MIX-TURE	APPROX. T.A.S.		TOT. GPH
			MPH	KTS	
2700	F.T.	RUN / RUN	360 / 385	310 / 335	80 / 97
2700	46	RUN / RUN	375 / 355	325 / 310	98 / 94
2700	46	RUN / RUN	355	310	103
2700	46	RUN / RUN	330 / 310 / 290	290 / 270 / 255	98 / 92 / 86

COLUMN II — SUBTRACT FUEL ALLOWANCES (4.05 STAT. (3.5 NAUT.) MI./GAL.)

RANGE IN AIRMILES			APPROX. T.A.S.		
STATUTE	NAUTICAL	R.P.M.	M.P. INCHES	MIX-TURE	MPH / KTS / TOT. GPH
1510 / 1440	1320 / 1260				
1370 / 1290	1190 / 1130				
1220 / 1150	1070 / 1000				
1080 / 1000 / 930	940 / 880 / 820				
		2600	44	RUN	97
		2600	44	RUN / RUN	350 / 325 / 305 / 305 / 280 / 90 / 85
		2600	44	RUN	305 / 245 / 80

COLUMN III — NOT AVAILABLE FOR CRUISING (4.5 STAT. (3.9 NAUT.) MI./GAL.)

RANGE IN AIRMILES					
STATUTE	NAUTICAL	R.P.M.	M.P. INCHES	MIX-TURE	TOT./MPH/KTS
1710 / 1630	1480 / 1410				
1550 / 1470	1340 / 1270				
1390 / 1310	1200 / 1130				
1220 / 1140 / 1060	1060 / 990 / 920				
		2500 / 2450 / 2350	F.T. / 43 / 43	RUN	89 / 84 / 79 / 365 / 340 / 325 / 315 / 295 / 280
		2300 / 2250	41 / 41	RUN	75 / 69 / 305 / 280 / 265 / 245
			40	RUN	64 / 280 / 225

COLUMN IV

RANGE IN AIRMILES		FUEL U.S. GAL.	PRESS ALT. FEET
STATUTE	NAUTICAL		
1910 / 1820	1660 / 1580	419 / 400	40000 / 35000 / 30000
1730 / 1640	1500 / 1430	380 / 360	25000
1550 / 1460	1350 / 1270	340 / 320	20000 / 15000
1370 / 1280 / 1190	1190 / 1110 / 1040	300 / 280 / 260	10000 / 5000 / S.L.

R.P.M.	M.P. INCHES	MIX-TURE	TOT. GPH	MPH	KTS
2500	F.T.	RUN	79	355	310
2300 / 2150	F.T. / 38	RUN / RUN / RUN	75 / 71 / 66	340 / 320 / 300	295 / 275 / 260
1950 / 1950	38 / 38	RUN / RUN	62 / 57	280 / 260	240 / 225
1950	37	RUN	53	240	205

COLUMN V

RANGE IN AIRMILES		
STATUTE	NAUTICAL	
2100 / 2010	1820 / 1740	
1810 / 1810	1650 / 1570	
1710 / 1610	1480 / 1400	
1510 / 1440 / 1320	1310 / 1230 / 1140	

MAXIMUM AIR RANGE

R.P.M.	M.P. INCHES	MIX-TURE	TOT. GPH	APPROX. T.A.S. MPH / KTS
2100 / 1850	F.T. / F.T.	RUN / RUN / RUN	60 / 58 / 54	295 / 285 / 265 / 255 / 245 / 230
1600 / 1600	34 / 33	RUN / RUN	49 / 42	240 / 220 / 205 / 210 / 190 / 175

LEGEND

ALT. = PRESSURE ALTITUDE	F.R. = FULL RICH		
M.P. = MANIFOLD PRESSURE	A.R. = AUTO-RICH		
GPH = U.S. GAL. PER HOUR	A.L. = AUTO-LEAN		
TAS = TRUE AIRSPEED	C.L. = CRUISING LEAN		
KTS. = KNOTS	M.L. = MANUAL LEAN		
S.L. = SEA LEVEL	F.T. = FULL THROTTLE		

EXAMPLE

At 10,500 lb. gross weight with 374 gal. of fuel (after deducting total allowances of 45 gal.) to fly 1600 stat. airmiles at 25,000 ft. altitude maintain 2300 RPM and F.T. in. manifold pressure with mixture set: RUN

SPECIAL NOTES

(1) Make allowance for warm-up, take-off & climb (see FIG. 53.) Plus allowance for wind, reserve and combat as required.

HIGH BLOWER ABOVE HEAVY LINE

DATA AS OF 9-10-44 BASED ON: FLIGHT TESTS

FLIGHT OPERATION INSTRUCTION CHART

AIRCRAFT MODEL(S): P-51D AND P-51K
ENGINE(S): V-1650-7
CHART WEIGHT LIMITS: 9600 TO 8000 POUNDS
EXTERNAL LOAD ITEMS: 2 – 75-GALLON COMBAT TANKS
NUMBER OF ENGINES OPERATING: 1

LIMITS	RPM	M.P. IN. HG.	BLOWER POSITION	MIXTURE POSITION	TIME LIMIT	CYL. TEMP.	TOTAL G.P.H.
WAR EMERG.	3000	67	LOW / HIGH	RUN / RUN	5 MIN.		211 / 215
MILITARY POWER	3000	61	LOW / HIGH	RUN / RUN	15 MIN.		182 / 187

INSTRUCTIONS FOR USING CHART: SELECT FIGURE IN FUEL COLUMN EQUAL TO OR LESS THAN AMOUNT OF FUEL TO BE USED FOR CRUISING. MOVE HORIZONTALLY TO RIGHT OR LEFT AND SELECT RANGE VALUE EQUAL TO OR GREATER THAN THE STATUTE OR NAUTICAL AIR MILES TO BE FLOWN. VERTICALLY BELOW AND OPPOSITE VALUE NEAREST DESIRED CRUISING ALTITUDE (ALT.) READ RPM, MANIFOLD PRESSURE (M.P.) AND MIXTURE SETTING REQUIRED.

NOTES: COLUMN I IS FOR EMERGENCY HIGH SPEED CRUISING ONLY. COLUMNS II, III, IV AND V GIVE PROGRESSIVE INCREASE IN RANGE AT A SACRIFICE IN SPEED. AIR MILES PER GALLON (MI./GAL) (NO WIND), GALLONS PER HR. (G.P.H.) AND TRUE AIRSPEED (T.A.S.) ARE APPROXIMATE VALUES FOR REFERENCE. RANGE VALUES ARE FOR AN AVERAGE AIRPLANE FLYING ALONE (NO WIND) TO OBTAIN BRITISH IMPERIAL GAL. (OR G.P.H.): MULTIPLY U.S. GAL. (OR G.P.H.) BY 10 THEN DIVIDE BY 12.

COLUMN I (3.6 STAT. / 3.15 NAUT. MI./GAL)

RANGE STATUTE	RANGE NAUTICAL	FUEL U.S. GAL.	PRESS ALT. FEET
870	790	280	40000
810	700	240	35000
740	640	220	30000
670	580	200	25000
600	520	180	20000
540	470	160	15000
470	410	140	10000
400	350	120	5000
340	290	100	S.L.
270	230	80	
200	170	60	
130	110	40	
60	60	20	

MAXIMUM CONTINUOUS (Column I)

R.P.M.	M.P. INCHES	MIX. TURE	TOT. GPH	T.A.S. MPH	T.A.S. KTS
2700	F.T.	RUN	90	370	320
2700	F.T.	RUN	97	390	335
2700	46	RUN	98	375	325
2700	46	RUN	94	355	310
2700	46	RUN	103	355	310
2700	46	RUN	98	335	290
2700	46	RUN	92	310	270
2700	46	RUN	86	290	255

COLUMN II (3.6 STAT. / 3.15 NAUT. MI./GAL)

RANGE STATUTE	RANGE NAUTICAL	R.P.M.	M.P. INCHES	MIX. TURE	TOT. GPH	T.A.S. MPH	T.A.S. KTS
970							
860							
790							
720							
640							
570							
500							
430							
360							
290							
200							
140							
70							
		2800	44	RUN	97		
		2600	44	RUN	90		
		2600	44	RUN	85		
		2600	44	RUN	80		

COLUMN III

Range/data – SUBTRACT FUEL ALLOWANCES NOT AVAILABLE FOR CRUISING:
850/750, 690/630/560, 500/440/380/310, 250/190/120/60

R.P.M.	M.P. INCHES	MIX.	GPH	MPH	KTS
2500	43	RUN	89	365	315
2450	42	RUN	82	340	295
2300	40	RUN	79	325	280
2300	F.T.	RUN			

COLUMN IV (4.1 STAT. / 3.55 NAUT. MI./GAL)

Range: 1100/980, 900/820/730, 650/570/490/410, 320/240/160/80

R.P.M.	M.P. INCHES	MIX.	GPH	MPH	KTS
2450	F.T.	RUN	77	355	310
2300	43	RUN	74	340	295
2100	F.T.	RUN	68	315	275
2150	F.T.	RUN	65	300	260
1900	37	RUN	59	275	240
1900	37	RUN	55	255	220
1850	37	RUN	51	235	205

COLUMN V (4.6 STAT. / 4.0 NAUT. MI./GAL)

STATUTE	NAUTICAL	FUEL U.S. GAL	PRESS ALT FEET
1370	1190	280	40000
1220	1060	240	35000
			30000
1120	970	220	25000
1010	880	200	20000
910	790	180	15000
810	700	160	10000
710	620	140	5000
610	530	120	S.L.
510	440	100	
400	350	80	
300	260	60	
200	170	40	
100	80	20	

MAXIMUM AIR RANGE

R.P.M.	M.P. INCHES	MIX. TURE	TOT. GPH	T.A.S. MPH	T.A.S. KTS
2200	29	F.T. RUN	60	305	265
2050	29	F.T. RUN	57	290	250
2050	F.T.	RUN	55	280	245
1750	F.T.	RUN	51	260	225
1600	32	RUN	47	240	210
1600	31	RUN	43	220	190
1600	30	RUN	39	200	175

SPECIAL NOTES

(1) MAKE ALLOWANCE FOR WARM-UP, TAKE-OFF & CLIMB (SEE FIG. 53) PLUS ALLOWANCE FOR WIND, RESERVE AND COMBAT AS REQUIRED.

HIGH BLOWER ABOVE HEAVY LINE

EXAMPLE

AT 9400 LB. GROSS WEIGHT WITH 230 GAL. OF FUEL (AFTER DEDUCTING TOTAL ALLOWANCES OF 30 GAL.) TO FLY 1000 STAT. AIRMILES AT 25,000 FT. ALTITUDE MAINTAIN 2300 RPM AND F.T. IN MANIFOLD PRESSURE WITH MIXTURE SET: **RUN**

LEGEND

- ALT. : PRESSURE ALTITUDE
- M.P. : MANIFOLD PRESSURE
- GPH : U.S. GAL. PER HOUR
- TAS : TRUE AIRSPEED
- KTS. : KNOTS
- S.L. : SEA LEVEL
- F.R. : FULL RICH
- A.R. : AUTO-RICH
- A.L. : AUTO-LEAN
- C.L. : CRUISING LEAN
- M.L. : MANUAL LEAN
- F.T. : FULL THROTTLE

DATA AS OF 9-10-44 BASED ON: FLIGHT TESTS

FLIGHT OPERATION INSTRUCTION CHART

AIRCRAFT MODEL(S): P-51D AND P-51K
ENGINE(S): V-1650-7
CHART WEIGHT LIMITS: 11,400 TO 10,000 POUNDS
EXTERNAL LOAD ITEMS: 2 - 110-GALLON COMBAT TANKS
NUMBER OF ENGINES OPERATING: 1

LIMITS	RPM	M.P. IN. HG.	BLOWER POSITION	MIXTURE POSITION	TIME LIMIT	CYL. TEMP.	TOTAL G.P.H.
WAR EMERG.	3000	67	LOW	RUN	5 MIN.		211
			HIGH	RUN			215
MILITARY POWER	3000	61	LOW	RUN	15 MIN.		182
			HIGH	RUN			187

For details see POWER PLANT CHART (FIG. 34, SECT. III)

COLUMN I (NAUT.) MI./GAL.

FUEL U.S. GAL.	RANGE IN AIRMILES STATUTE	NAUTICAL
489	1620	1400
460	1520	1320
440	1450	1260
420	1380	1200
400	1320	1150
380	1250	1090
360	1190	1030
340	1120	970
320	1060	920
300	990	860
280	920	800
260	860	740

MAXIMUM CONTINUOUS

R.P.M.	M.P. INCHES	MIXTURE	TOT. GPH	T.A.S. MPH	KTS
2700	F.T.	RUN	80	345	300
2700	F.T.	RUN	97	380	330
2700	46	RUN	98	370	320
2700	46	RUN	103	385	335
2700	46	RUN	108	350	305
2700	46	RUN	98	330	285
2700	48	RUN	92	310	270
2700	46	RUN	86	290	250

COLUMN II (3.3 NAUT.) MI./GAL.

(Blank - SUBTRACT FUEL ALLOWANCES NOT AVAILABLE FOR CRUISING)

COLUMN III (3.8 STAT. (3.3 NAUT.) MI./GAL.)

RANGE IN AIRMILES STATUTE	NAUTICAL
1900	1650
1790	1550
1720	1490
1640	1420
1570	1350
1490	1290
1410	1220
1340	1160
1260	1090
1190	1030
1110	960
1030	890

R.P.M.	M.P. INCHES	MIXTURE	TOT. GPH	T.A.S. MPH	KTS
2500	F.T.	RUN	88	335	290
2450	42	RUN	81	310	270
2450	42	RUN	76	290	250
2400	42	RUN	71	270	235

COLUMN IV (4.3 STAT. (3.75 NAUT.) MI./GAL.)

RANGE IN AIRMILES STATUTE	NAUTICAL
2130	1860
2010	1750
1920	1670
1840	1600
1750	1520
1660	1450
1580	1370
1490	1300
1410	1220
1320	1150
1230	1070
1150	1000

R.P.M.	M.P. INCHES	MIXTURE	TOT. GPH	T.A.S. MPH	KTS
2350	F.T.	RUN	79	340	295
2250	40	RUN	75	320	280
2100	39	RUN	71	305	265
2100	39	RUN	66	285	245
2050	39	RUN	62	265	230
		RUN	57	245	215

COLUMN V

FUEL U.S. GAL.	RANGE IN AIRMILES STATUTE	NAUTICAL
489	2360	2050
460	2220	1930
440	2120	1850
420	2030	1770
400	1940	1680
380	1840	1600
360	1750	1520
340	1650	1440
320	1560	1350
300	1460	1270
280	1370	1190
260	1270	1110

PRESS ALT. FEET
40000
35000
30000
25000
20000
15000
10000
5000
S.L.

MAXIMUM AIR RANGE

R.P.M.	M.P. INCHES	MIXTURE	TOT. GPH	T.A.S. MPH	KTS
2150	F.T.	RUN	59	280	245
1950	F.T.	RUN	57	270	235
1650	35	RUN	52	245	215
1600	34	RUN	48	225	195
1600	33	RUN	43	205	180

NOTES: COLUMN I IS FOR EMERGENCY HIGH SPEED CRUISING ONLY. COLUMNS II, III, IV AND V GIVE PROGRESSIVE INCREASE IN RANGE AT A SACRIFICE IN SPEED. AIR MILES PER GALLON (MI./GAL.) AND NO WIND, GALLONS PER HR. (G.P.H.) AND TRUE AIRSPEED (T.A.S.) ARE APPROXIMATE VALUES FOR REFERENCE. RANGE VALUES ARE FOR AN AVERAGE AIRPLANE FLYING ALONE (NO WIND). TO OBTAIN BRITISH IMPERIAL GAL. (G.P.H.): MULTIPLY U.S.GAL (OR G.P.H.) BY 10 THEN DIVIDE BY 12.

INSTRUCTIONS FOR USING CHART: SELECT FIGURE IN FUEL COLUMN EQUAL TO OR LESS THAN AMOUNT OF FUEL TO BE USED FOR CRUISING. MOVE HORIZONTALLY TO RIGHT OR LEFT AND SELECT RANGE VALUE EQUAL TO OR GREATER THAN THE STATUTE OR NAUTICAL AIR MILES TO BE FLOWN. VERTICALLY BELOW AND OPPOSITE VALUE NEAREST DESIRED CRUISING ALTITUDE (ALT.) READ RPM, MANIFOLD PRESSURE (M.P.) AND MIXTURE SETTING REQUIRED.

SPECIAL NOTES
(1) MAKE ALLOWANCE FOR WARM-UP, TAKE-OFF & CLIMB (SEE FIG. 53) PLUS ALLOWANCE FOR WIND, RESERVE AND COMBAT AS REQUIRED.

HIGH BLOWER ABOVE HEAVY LINE

EXAMPLE
AT 11,200 LB. GROSS WEIGHT WITH 440 GAL. OF FUEL (AFTER DEDUCTING TOTAL ALLOWANCES OF 140 GAL.) TO FLY 1900 STAT. AIRMILES AT 25,000 FT. ALTITUDE MAINTAIN 2350 RPM AND F.T. IN. MANIFOLD PRESSURE WITH MIXTURE SET: RUN

DATA AS OF 9-10-44 BASED ON: FLIGHT TESTS

LEGEND

ALT.	= PRESSURE ALTITUDE	F.R.	= FULL RICH
M.P.	= MANIFOLD PRESSURE	A.R.	= AUTO-RICH
GPH	= U.S. GAL. PER HOUR	A.L.	= AUTO-LEAN
TAS	= TRUE AIRSPEED	C.L.	= CRUISING LEAN
KTS	= KNOTS	M.L.	= MANUAL LEAN
S.L.	= SEA LEVEL	F.T.	= FULL THROTTLE

FLIGHT OPERATION INSTRUCTION CHART

AIRCRAFT MODEL(S): P-51D AND P-51K
ENGINE(S): V-1650-7

CHART WEIGHT LIMITS: 10,000 TO 8500 POUNDS

EXTERNAL LOAD ITEMS: 2 – 110-GALLON COMBAT TANKS
NUMBER OF ENGINES OPERATING: 1

LIMITS	R.P.M.	M.P. IN.HG.	BLOWER POSITION	MIXTURE POSITION	TIME LIMIT	CYL. TEMP.	TOTAL G.P.H.
WAR EMERG.	3000	67	LOW	RUN	5 MIN.		211
			HIGH	RUN	5 MIN.		215
MILITARY POWER	3000	61	LOW	RUN	15 MIN.		182
			HIGH	RUN	15 MIN.		187

For details see Power Plant Chart (Fig. 34, Sect. III)

NOTES: COLUMN I IS FOR EMERGENCY HIGH SPEED CRUISING ONLY. COLUMNS II, III, IV AND V GIVE PROGRESSIVE INCREASE IN RANGE AT A SACRIFICE IN SPEED. AIR MILES PER GALLON (MI./GAL.) (NO WIND), GALLONS PER HR. (G.P.H.) AND TRUE AIRSPEED (T.A.S.) ARE APPROXIMATE VALUES FOR REFERENCE. RANGE VALUES ARE FOR AN AVERAGE AIRPLANE FLYING ALONE (NO WIND)[1]. TO OBTAIN BRITISH IMPERIAL GAL. (OR G.P.H.) MULTIPLY U.S. GAL. (OR G.P.H.) BY 10 THEN DIVIDE BY 12.

INSTRUCTIONS FOR USING CHART: SELECT FIGURE IN FUEL COLUMN EQUAL TO OR LESS THAN AMOUNT OF FUEL TO BE USED FOR CRUISING. MOVE HORIZONTALLY TO RIGHT OR LEFT AND SELECT RANGE VALUE EQUAL TO OR GREATER THAN THE STATUTE OR NAUTICAL AIR MILES TO BE FLOWN. VERTICALLY BELOW AND OPPOSITE VALUE NEAREST DESIRED CRUISING ALTITUDE (ALT.) READ RPM, MANIFOLD PRESSURE (M.P.) AND MIXTURE SETTING REQUIRED.

COLUMN I (3.6 STAT. / 3.15 NAUT.) MI./GAL.

RANGE IN AIRMILES		FUEL	APPROX. T.A.S.			
STATUTE	NAUTICAL	U.S. GAL.	TOT. GPH	MPH	KTS	
890 / 790	770 / 690	269 / 240	80 / 97	365 / 385	315 / 335	
720 / 660 / 580	630 / 570 / 510	220 / 200 / 180	98 / 94 / 103	370 / 350 / 350	320 / 305 / 305	
530 / 460 / 330	460 / 400 / 280	160 / 140 / 100				
260 / 200 / 130	230 / 170 / 110	80 / 60 / 40				
	50	20				

MAXIMUM CONTINUOUS

R.P.M.	M.P. INCHES	MIXTURE
2700 / 2700	F.T. / F.T.	RUN / RUN
2700 / 2700 / 2700	46 / 46 / 46	RUN / RUN / RUN
2700 / 2700 / 2700	46 / 46 / 46	RUN / RUN / RUN

COLUMN II (3.6 STAT. / 3.15 NAUT.) MI./GAL.

RANGE IN AIRMILES			APPROX. T.A.S.		
STATUTE	NAUTICAL	TOT. GPH	MPH	KTS	
970 / 880	850 / 750				
790 / 720 / 650	690 / 630 / 560				
570 / 500 / 430 / 360	500 / 440 / 380 / 310				
280 / 210 / 140 / 70	250 / 190 / 120 / 60				

R.P.M.	M.P. INCHES	MIXTURE	TOT. GPH	MPH	KTS
2550	44	RUN	94	340	295
2550 / 2550 / 2500	44 / 44 / 43	RUN / RUN / RUN	88 / 81 / 76	320 / 295 / 275	275 / 255 / 240

COLUMN III (4 STAT. / 3.45 NAUT.) MI./GAL.

SUBTRACT FUEL ALLOWANCES NOT AVAILABLE FOR CRUISING[1]

RANGE IN AIRMILES			APPROX. T.A.S.		
STATUTE	NAUTICAL	TOT. GPH	MPH	KTS	
1070 / 960	930 / 830				
880 / 800 / 720	760 / 690 / 620				
640 / 560 / 480 / 400	550 / 480 / 410 / 340				
320 / 240 / 160 / 80	270 / 200 / 130 / 70				

R.P.M.	M.P. INCHES	MIXTURE	TOT. GPH	MPH	KTS
2500 / 2500 / 2400	43 / 43 / F.T.	RUN / RUN / RUN	90 / 85 / 82	365 / 340 / 325	315 / 295 / 280
2300 / 2300 / 2300	41 / 41 / 41	RUN / RUN / RUN	75 / 70 / 65	300 / 280 / 260	260 / 245 / 225

COLUMN IV (4.4 STAT. / 3.85 NAUT.) MI./GAL.

RANGE IN AIRMILES		FUEL	APPROX. T.A.S.		
STATUTE	NAUTICAL	U.S. GAL.	TOT. GPH	MPH	KTS
1180 / 1050	1030 / 920	269 / 240			
960 / 880 / 790	840 / 770 / 690	220 / 200 / 180			
700 / 610 / 520 / 440	610 / 540 / 460 / 380	160 / 140 / 120 / 100			
350 / 280 / 170 / 90	310 / 230 / 150 / 70	80 / 60 / 40 / 20			

R.P.M.	M.P. INCHES	MIXTURE	TOT. GPH	MPH	KTS
2500	F.T.	RUN	82	360	310
2300 / 2150 / 2200	F.T. / F.T. / 38	RUN / RUN / RUN	77 / 71 / 68	340 / 315 / 300	295 / 275 / 260
2000 / 2000 / 2000	38 / 38 / 38	RUN / RUN / RUN	63 / 59 / 55	280 / 260 / 240	245 / 225 / 210

COLUMN V – MAXIMUM AIR RANGE

RANGE IN AIRMILES		FUEL	PRESS ALT. FEET	APPROX.		
STATUTE	NAUTICAL	U.S. GAL.				
1330 / 1190	1150 / 1030	269 / 240	40000 / 35000 / 30000			
1090 / 990 / 890	940 / 860 / 770	220 / 200 / 180	25000 / 20000 / 15000			
790 / 690 / 590 / 490	690 / 600 / 510 / 430	160 / 140 / 120 / 100	10000 / 5000 / S.L.			
390 / 290 / 200 / 100	340 / 250 / 170 / 80	80 / 60 / 40 / 20				

R.P.M.	M.P. INCHES	MIXTURE	TOT. GPH	MPH	KTS
2050 / 2100 / 1800	F.T. / F.T. / F.T.	RUN / RUN / RUN	59 / 56 / 52	290 / 280 / 260	250 / 240 / 225
1650 / 1600 / 1600	32 / 32 / 31	RUN / RUN / RUN	49 / 45 / 41	240 / 220 / 200	210 / 190 / 175

SPECIAL NOTES

(1) MAKE ALLOWANCE FOR WARM-UP, TAKE-OFF & CLIMB (SEE FIG. 53) PLUS ALLOWANCE FOR WIND, RESERVE AND COMBAT AS REQUIRED.

HIGH BLOWER ABOVE HEAVY LINE

EXAMPLE

AT 9800 LB. GROSS WEIGHT WITH 230 GAL. OF FUEL (AFTER DEDUCTING TOTAL ALLOWANCES OF 39 GAL.) TO FLY 900 STAT. AIRMILES AT 25,000 FT. ALTITUDE MAINTAIN 2300 RPM AND F.T. IN. MANIFOLD PRESSURE WITH MIXTURE SET: RUN

LEGEND

ALT. = PRESSURE ALTITUDE
M.P. = MANIFOLD PRESSURE
GPH = U.S. GAL. PER HOUR
TAS = TRUE AIRSPEED
KTS. = KNOTS
S.L. = SEA LEVEL
F.R. = FULL RICH
A.R. = AUTO-RICH
A.L. = AUTO-LEAN
C.L. = CRUISING LEAN
M.L. = MANUAL LEAN
F.T. = FULL THROTTLE

DATA AS OF 9-10-44   BASED ON: FLIGHT TESTS

INDEX

	Page
Acrobatics	78
After-cooling system	24
Ailerons, description	8
Airspeed limitations	71
Ammeter	19, 83
Armament	40
Armor	46
Bailout	87
In fire	85
In spin	88
Over water	88
Belly landing procedure	80
Bombing equipment	44
Brakes, operation	18
Failure of	82
Camera, gun	41
Canopy	25
Carburetor	13
Checks, before starting	50
External	49
Oxygen	39
Pre-takeoff	54
Chemical tanks	45
Circuit breakers	19, 83
Cockpit	26
Cockpit arrangement	9, 10, 11
Compressibility	72
Recovery from	75
Control lock	8
Controls	8
Coolant system	24
Failure of	81
Data case	48
Dimensions	7
Ditching procedure	86
Dives, high speed	68
Compressibility	72
Recovery procedure	70
Uncontrolled	74
Drop message bag	48
Electrical system	19
Failure of	83
Lights	92
Emergency procedures	79
Bailout	87
Belly landing	80
Brakes, failure of	82
Coolant system, failure of	81
Ditching	86
Electrical system, failure of	83
Engine overheating	81
Fire	85
Hydraulic system, failure of	82
Oil system, failure of	81
Radio emergency procedure	36
Runaway propeller	81
Tire failure	84
Engine	12
Instrument limits	52
Overheating of	81
Fire, bailout procedure	85
Flaps	8
Flare gun	47
Flight characteristics	65
Gunnery	43
High altitude	67
With droppable tanks	67
With fuselage tank	67
Flight operation charts, use of	94
Flight operation instruction charts	97
Forced landings	79
Fuel system	20
Gliding	76
Go-around procedure	64
Gun camera	41
Guns	40
Gunsight	42
Gunnery, effects on flight	43
High altitude characteristics	67
High-speed diving	68
Homing	34
Hydraulic system	18
Failure of	82
Instruments	27
Limits	52, 66
Instrument flying	89

104

	Page
Landing procedure	56
Belly	80
Common errors in	60
Crosswind	62
Forced	79
Gusty	63
Tips on	62
Wet field	63
Landing gear	16
Emergency operation	82
Warning system	17
Lock, control	8
Mach number	73
Magnetos, checking	55
Manifold pressure regulator	13
Manifold pressures, limits	52
Map case	48
Miscellaneous equipment	48
Night flying	92
Forced landing at night	81
Oil dilution	24
Oil system	23
Failure of	81
Overheating, engine	81
Oxygen system	23
Check	39
Parking brakes	18

	Page
Propeller	16
Runaway	81
Radio	31
Emergency procedure	36
Homing	34
Navigation	34
Recognition lights	47
Relief tube	48
Reversibility	67
Rockets	43
Shoulder harness	26
Signalling equipment	47
Speeds, maximum allowable	71
True versus indicated airspeed	73
Spins	77
Stalls	76
Starting the engine	51
Stopping the engine	52
Supercharger	13, 68
Takeoff, climb, and landing chart	96
Takeoff procedure	55
Taxiing	53
Throttle quadrant	15
Tie-down kit	48
Tire failure	84
Ventilation, cockpit	26
War emergency power	14

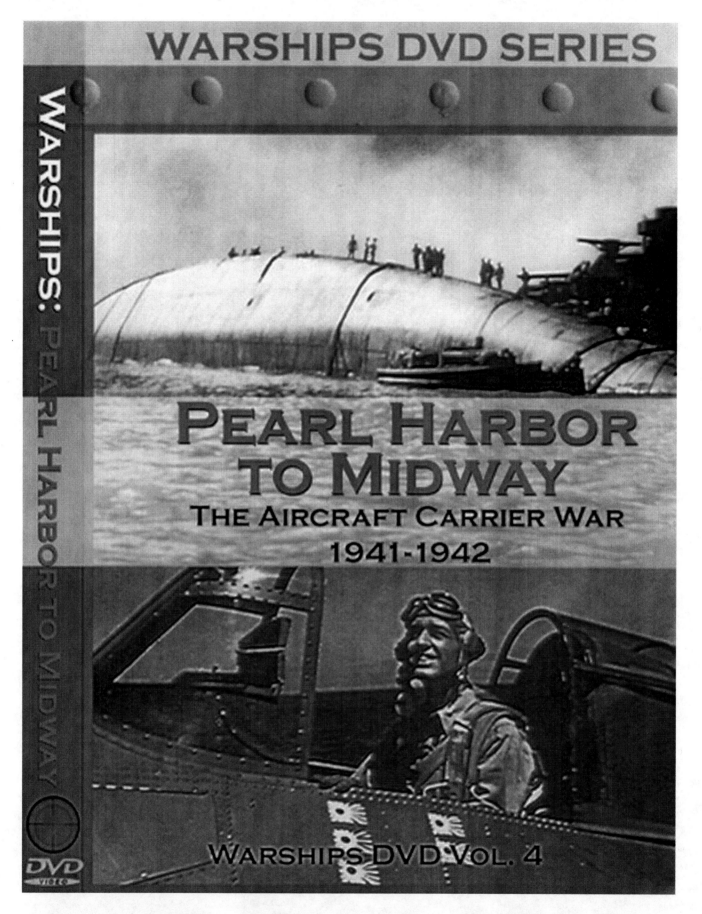

Epic Battles of WWII

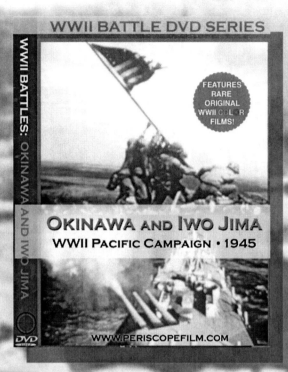

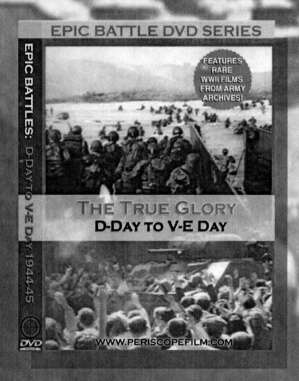

Now Available on DVD!

ALSO NOW AVAILABLE FROM PERISCOPEFILM.COM

©2006-2009 Periscope Film LLC
All Rights Reserved
ISBN #978-1-4116-9040-0

CPSIA information can be obtained at www.ICGtesting.com
Printed in the USA
BVOW04s1811010615

402325BV00023B/118/P